Stars and Telescopes for the Beginner

To Kevin

Stars and Telescopes for the Beginner

By Roy Worvill

Illustrations by Stephen Weston

Taplinger Publishing Company, New York

First published in the United States in 1979 by
TAPLINGER PUBLISHING CO., INC.
New York, New York

Library of Congress Catalog Card Number: 78-68766
ISBN 0-8008-7368-8

Grateful acknowledgement is made by the author to
Stephen Weston for the drawings in this book and to
Morris Kahn, of Kahn & Averill, for his helpful
suggestions.

Contents

*All is peculiar, all is magical   We look into a rose and discover
a universe; we contemplate a butterfly and come upon a miracle.
Every grass-blade is a cry to heaven, every dust-mote a whirling
planet. To reduce truth to the rational and the logical is to deny
truth. Truth is a curlew that gives its clear call; the light flashes,
a feather drops at our feet and it has flown away.*

Alyse Gregory
The Day Is Gone

*What a thrilling time you must be having with that telescope, looking
at the stars. There is nothing, not anything, so exciting.*

From The Letters of Edna St.Vincent Millay.

Foreword

The popular image of an astronomer, as represented by a venerable, bearded gentleman seated at a telescope, with one eye on heaven and the other on his observatory clock or voluminous notebook, is no longer valid. The beard may still be present, but its wearer is more likely perched in a cage, high above ground level, inside an immense structure which is scarcely recognisable as a telescope, conducting a photographic reconnaissance of remote galaxies whose light has taken many millions of years to reach the earth. If his field of research lies in radio astronomy he may be watching the peaks and troughs, like those of a temperature chart, being etched by a pen on a moving paper scroll, the visible signature of radio waves coming from the expanding gas cloud of a star which exploded with catastrophic force before human eyes first blinked at the sunlight, or human minds began to grapple with the baffling mystery of the universe around them.

The instrumental armoury which now lies at the disposal of astronomers is, indeed, formidable. On our television screens we have watched human footprints being impressed in the dusty soil of the moon and seen, at close range, the boulder-strewn, orange deserts and craters of the planet Mars. Even the amateur astronomer's telescope has advanced in size, sophistication and refinement, far beyond the simple 'optic tube' which first wrought its wonders to the eye of Galileo in the winter of 1609-10. It is no longer a narrow cylinder enclosed by a pair of spectacle lenses, though so primitive an implement is still not to be despised, and the words of W.F. Denning, written many years ago in his delightful old book, *Telescopic Work for Starlight Evenings* yet bear the authentic ring of truth, in spite of space-probes, spy-satellites, computer technology and giant radio dishes.

'Even to him who simply makes the science a hobby and a source of recreation in a leisure hour after the cares of business, the sky never ceases to afford a means of agreeable entertainment. The observer who, quietly, from his cottage window, surveys the evening star or the new moon through his little telescope, often finds a deeper pleasure than the professional astronomer, who, from his elevated and richly-appointed observatory, discovers new orbs with one of the most powerful instruments ever made'.

However, the cottage window is far from an ideal observing site and the amateur's opportunities are by no means limited to the simple star-gazing conjured up by Denning's idyllic picture. Indeed, Denning himself made valuable contributions to serious research on the planets, comet discoveries and meteors. His instrumental resources were in no way superior to those which are still readily available to any amateur today, a ten-inch Newtonian reflecting telescope, mounted outdoors on a common altazimuth tripod, while the atmospheric conditions at Bristol were often discouraging. The chances for the amateur to make important discoveries may be fewer than they were in Denning's time, but the delights and the mystery remain undiminished. The rewards, as Emerson once said, are rich and royal.

Roy Worvill
Chipping Norton, Oxon.
February 1979

Glossary of Technical Terms

Aberration   faults which are inherent in every optical system to greater or lesser degrees. The most common forms of aberration are described individually below.

Achromatic   means free from the coloured fringes produced by the refraction of light (chromatic aberration). Though formerly applied to any refracting telescope having a compound object-glass, it is the reflecting form which is truly achromatic. Even the three-element photo-visual objective cannot produce perfect achromatism although its residual colour is minimal.

Airy disc   the central disc of a star-image seen through a telescope. It is surrounded by a series of concentric rings produced by diffraction effects in the converging pencil of light; otherwise known as the spurious disc; named after the former Astronomer Royal, Sir George Airy (1801-1892).

Altazimuth   simple form of telescope mounting which provides for motion in the vertical and horizontal planes (altitude and azimuth).

Anastigmat   a lens which has been corrected for the aberration of astigmatism (See below). Modern photographic anastigmats are manufactured in a great variety of complex designs. One of the best-known simple forms is the four-element Zeiss Tessar which has been widely copied.

Aplanat   a type of object-glass which has been corrected for the faults of coma and spherical aberration (See below)

Apochromat   a highly-corrected design of object-glass, usually a triplet combination of lenses, which closely approaches perfect achromatism. The Cooke photovisual is an example.

Astigmatism   an objectionable form of aberration which arises when a lens or mirror is not a surface of revolution. If, for example, the rays of light from the ends of one diameter of an

9

object-glass are refracted (or reflected by a mirror) to a shorter focus than those from a diameter at right-angles to it, there will be no well-defined image produced in a single plane.

Barlow lens　a negative (diverging) lens which is placed in the converging cone of light from a lens of mirror. By decreasing the convergence of the cone it extends the effective focal length of the objective and thereby increases the overall magnification. Most take the form of a plano-concave doublet.

Berthon dynamometer　a device to determine the magnification of an eyepiece used with a given objective, whether a lens or mirror. It consists of a graduated scale with a V-shaped slot against which the column of light emerging from the eyepiece, (the exit pupil) is measured. If the diameter of the objective is divided by the diameter of the exit pupil, the resulting quotient is the magnifying power of the eyepiece with that particular objective.

Catadioptric　a form of telescope which combines a lens and a mirror in producing an image at the focus. The Maksutov design is an example of a compact and highly-efficient catadioptric telescope.

Chromatic aberration　the troublesome colour fault arising from the inability of a simple lens to bring light of all colours to the same focus. The 'achromatic' doublet corrects this only partially.

Collimation　the correct alignment of the optical elements in a telescope or pair of binoculars. Errors of collimation will have an adverse effect upon the performance of even the finest instrument. Accurate collimation of a telescope is not usually difficult, but binocular collimation is strictly for the professional. Film and television producers have not, so far, learned that a correctly-collimated pair of binoculars gives a circular field of view like that of a telescope. Their cameras may be efficient but their binoculars invariably give a field of view shaped like a pair of spectacles!

Coma　a form of aberration which makes a lens or the mirror of the reflecting telescope, especially if of short focal length, incapable of producing good definition over a wide field. A star may be perfectly-defined on the optical axis, at the centre of the field, but away from the axis the images become increasingly fan-shaped. The Schmidt telescope is one well-known and effective design which overcomes this.

Coronagraph　　an instrument designed by Lyot in 1930 which enables the sun's corona to be studied at any time when the sun is visible. Before this discovery such observations were restricted to the rare and fleeting moments of a total solar eclipse.

Coudé mounting　　a design of telescope mounting which provides observation from a fixed position instead of the normal moving eyepiece. This is done by means of a reflecting element, a prism or plane mirror, which moves and brings the image to the point of observation.

Dawes Limit　　takes its name from a celebrated amateur observer, the Rev. W.R. Dawes, who carried out an exhaustive series of tests to determine the resolving powers of telescopes, which means their ability to separate point sources of light such as double stars. He concluded that a telescope of 1-inch (25 mm) aperture would separate double stars if the components were not closer than 4.56 seconds of arc. The resolving powers of larger apertures are greater and are determined by dividing 4.56 by the diameter of the objective. A lens or mirror of 4½ inches aperture should, therefore, be capable of resolution to approximately 1 second of arc. In practice the limit is subject to some variable factors such as the state of the atmosphere and the observer's eye.

Declination　　the angular distance of an object north or south of the celestial equator measured on the great circle passing through the pole and the object.

Diffraction　　the effect produced by light shining through a hole or slit where the waves are bent at the edge and interfere with each other. This results in a pattern of alternating light and dark bands seen in the out-of-focus view of a star and also by observing a street lamp through the woven fabric of an open umbrella.

Equation of Time　　the difference between the time shown on a sun-dial and Greenwich Mean Time. The sun's apparent motion is not uniform in speed due to the elliptical shape of the earth's orbit and the inclination of the sun's path, the ecliptic, to the equator. The two agree on only four days in the year. The difference may be as much as 17 minutes.

Exit pupil   the column of light emerging from the eyepiece of a telescope or from each part of a pair of binoculars if the instrument is pointed to a window or the sky. A 7 x 50 pair of binoculars has an exit pupil of just over 7 mm diameter (50 divided by 7) which is larger than the normal eye can accommodate in full daylight. In twilight, with the eye pupil wide open, the advantage of the large exit pupil is appreciable.

Field curvature   an aberration resulting from the tendency of light rays which strike the objective obliquely to come to a shorter focus than those which pass along the axis. The effect is not serious in telescopes of normal focal ratio such as f/8 for a mirror and f/15 for an object-glass.

Flexure   the slight bending, with consequent distortion of figure and harmful effect upon image quality, which arises when an object-glass or mirror is not adequately supported in its cell. A large mirror can be given support at a number of points but the lens can be held only at the edge, which is one of the factors limiting the size of object-glasses in use.

Focal ratio   or f/ number is used in telescope objectives as it is with photographic lenses and relates the effective diameter of the lens or mirror to its focal length. A 6-inch lens with a focal length of 60 inches is said to work at f/10. In the case of a 'fast' camera lens the fractional value of the figure may be higher, such as f/2 or f/1.4.

Fork Mounting   a very convenient and efficient form of telescope mount which supports the tube on short trunnion arms rotating in the two arms of the fork. If the main axis is vertical it is altazimuth in form but normally the axis is tilted to the north celestial pole to give equatorial motion.

German equatorial   a form of mounting devised by Fraunhofer (1789-1826). The telescope is supported at one end of the declination axis, at the other end of which it is counter-balanced by a weight.

Ghost image   a false image produced in some forms of eyepiece by internal reflections. The effect can be disconcerting and misleading if the observer is looking for a very faint object. The monocentric eyepiece is a form which enjoys complete freedom from such 'ghosts'.

Herschelian telescope   as its name implies it was the form of reflector used by Sir William Herschel. The image was brought to the observer's eye at the open end of the tube by tilting the main mirror. The aim was to avoid a second reflection such as the Newtonian flat produces, and which, in the days of metal mirrors, involved a serious loss of light.

Horseshoe mounting   a form of mount devised by Russell Porter for the 200-inch Palomar reflector; not suitable for refractors or reflectors of the Cassegrain type.

Light-grasp   is the function of the objective, whether lens or mirror. It varies as the square of the diameter. By doubling the diameter we quadruple the area and hence the light-grasp. More light brings fainter objects into view. The light-grasp of the reflector's mirror system is obviously reduced by a tarnished silver or aluminium coating, as is that of the refractor by the accumulation of dust or dew on the lens.

Meniscus   a form of lens with one surface convex and the other concave. The Maksutov telescope employs a steeply-curved meniscus as the corrector plate or 'shell'.

Objective   the image-forming element of a telescope; the object-glass in the refractor and the main mirror of a reflector.

Ocular   an alternative name for a telescope eyepiece.

Parallax   the apparent movement of an object arising from the real motion of the observer. Observation of a near object by each eye in succession produces the most familiar example of parallax. The object appears to move its position slightly, relative to more distant ones.

Personal equation   the human factor which can never be entirely eliminated from observational data and which inevitably involves some margin of error in timing, recording or interpretation of phenomena.

Photovisual   a type of object-glass corrected for both visual and photographic observation. (See apochromat above)

Pyrex   a valuable type of glass widely-used for mirror-making.

Its chief virtue is that of low-expansion, since temperature variations can affect the 'figure' of a mirror and therefore its performance. The advantage is more important in large mirrors.

Ramsden disc   an alternative name for exit pupil (See above)

Rayleigh limit   the limit of tolerance for errors in the curvature of a lens or mirror objective. It is expressed in terms of the wavelength of light. For satisfactory performance it is usually accepted that the mirror should have an accuracy of 1/8-wave. For a refractor the tolerance is less stringent since the light passes only once along the tube. In the reflector, where it travels twice along the tube, the effect of an error of curvature is doubled.

Resolution   the ability to show detail in an extended image such as that of the moon or a planet and to separate the components of double stars. Resolving power increases directly with the aperture of a telescope but is also affected by atmospheric conditions and the observer's eyesight. (See Dawes' Limit above)

Schmidt Camera   Designed by Bernard Schmidt in 1930. Uses a spherical mirror with a lens or correcting plate to overcome the aberrations produced by the mirror. It is especially effective as a photographic instrument because it combines fast speed with a very wide field of cover, though the plate or film must be shaped to the correct curve.

Secondary spectrum   the coloured fringe of light which is not brought to a focus by the ordinary two-element object-glass. In a small instrument it is not obtrusive but is liable to become so in larger apertures, above 6 inches. Light-grasp, resolution and contrast in planetary images are also affected to some extent compared with the perfect achromatism of the mirror.

Speculum   an old name for a telescope mirror. The first mirrors were made of 'speculum metal', the composition of which varied probably, for the early opticians were secretive about their craft. Usually an alloy of copper and tin and has been called by Bell ('The Telescope'), 'a peculiarly mean metal to cast and work'.

Spherical aberration   produced by any lens or mirror of spherical surface when light-rays refracted by, or reflected from such a surface are not brought to a common focus. Rays from the outer zone reach a focus nearer to the lens or mirror than the centre rays.

Spider   the name applied to the supports of the flat mirror in the Newtonian reflector. The support may be a single arm or consist of three or four members. Another design which has been recommended consists of two interlocking circles.

Springfield mounting   has proved popular in the United States where it originated. It provides a fixed observing position for the user of a Newtonian, but involves additional reflecting surfaces and a 'right-left' inversion of star-fields and extended images such as those of the moon and planets.

Spurious disc   the small disc of a star seen when the telescope is focussed on it. Its size decreases with larger apertures. (See Airy disc above)

Steadying rod   a device for reducing vibrations of the telescope and giving greater rigidity, usually by a series of rods like a toasting-fork attaching the eyepiece end of the tube to the tripod in a refractor, or a stout single rod combining the slow motion control in altitude, for the Newtonian.

Sun diagonal   a form of eyepiece attachment which gives a reflected image of the sun for magnification while allowing the heat to pass out of the telescope without reaching the observer's eye. A dark glass screen is still needed for use with the eyepiece. Better still is to avoid direct observation of the sun entirely and project the sun's image on a screen of white card.

Zonal aberration   a form of optical error familiar to amateur makers of mirrors. It arises from irregularities in the figure which produce two or more concentric zones of varying focal length.

1 The Sky Around Us

There are few things more deceptive than the apparent stillness of the solid earth beneath our feet. 'Down to earth' is a saying which has long been associated with hard fact and what we have been taught to think of as the basic truth. It can hardly surprise us that our ancestors relinquished, with extreme reluctance, the illusion of a stationary Earth at the centre of the universe; for all appearances, not to mention their instinctive self-esteem, seemed to support their convictions as to the position and status of their world. To uproot the belief in an Earth-centred universe and replace it with the modern, scientific view, took centuries of bitter controversy. Even after the publication in the year 1543 of the famous book by Copernicus, *On the Revolution of the Celestial Orbs*, dedicated to the Pope and its printing paid for by a cardinal of the church, the general acceptance of such a revolutionary theory was not easily accomplished. Indeed more than a half-century after its publication, in 1616, the book was declared heretical, 'false and altogether contrary to Holy Scripture'. As he received a copy of the book on his death-bed, Copernicus was under no misapprehensions as to its probable reception in some quarters. 'If', he declared, 'there are some babblers who, though ignorant of mathematics, take it upon themselves to judge these matters and dare to criticise my work because of some passage of Scripture which they have wrested to their own purpose, I regard them not and I will not scruple to hold their judgement in contempt'.

The evidence which was produced by the first telescopic observations of Galileo in support of Copernicus was treated with scant respect, but the gradual accumulation of observational data, together with the work of Johannes Kepler in establishing the laws which governed the movements of planets and Isaac Newton's formulation of celestial mechanics a few years later, completed the inexorable ascendancy of the new view of the universe. It is interesting to reflect that the passing of another three centuries has scarcely modified the views of Kepler, Galileo and Newton — except in respect of bodies moving at speeds comparable to that of light. It is only in these rare instances that Einstein's theory of relativity has proved capable of explaining certain phenomena, such as the slow changes taking place in the orbit of the planet Mercury, which are not readily explicable in terms of Newtonian mechanics. By gradual, and often painful, stages in the advance of scientific knowledge, we have exchanged the solidly-based, Earth-centred universe of our forefathers, for life on a moving space-craft, spinning round its axis every 23 hours 56 minutes and 4 seconds, following an orbit round the sun which it completes in 365.24 days and taking part with the sun and the other planets in the revolution of the vast spiral galaxy in which the sun is but one medium-sized star among many thousands of millions. In the rest of this chapter we shall consider the way in which all these motions are reflected in the view we enjoy of the sky by day and night. In succeeding chapters we are concerned with the instruments and methods used to achieve a closer look at some of our fellow-travellers in space. It is sometimes said that the lone traveller is able to travel fastest, but in our journey through space we have no choice. The Earth's orbital velocity of 18.5 miles every second should be fast enough to satisfy even today's love of speed, but it is, on the other hand, a pace which is still leisurely enough for us to enjoy the changing panorama of the scenery, whether we use a pair of binoculars, a telescope or the naked eye.

Mapping the Sky

The extensive range of instruments which the modern astronomer has at his command in a great observatory, to say nothing of the astonishing and complex apparatus associated with recent spectacular space probes to the moon and planets, may easily cause the beginner in astronomy to overlook the achievements of the pre-telescopic era. Even the observers of the ancient world, whose lives ante-date the telescope by more than a thousand years, had some remarkable achievements to their credit. The Greek philosopher, Thales of Miletus, was certainly among the earliest astronomers. He is said to have successfully predicted an eclipse of the sun in 585 B.C. probably by making use of the cycle known as the saros which had been discovered by the Babylonians long before his time. This interval of 223 lunar months, covering a period of 18 years 11½ days, produces cycles of eclipses of marked similarity.

Another famous Greek astronomer was Aristarchus of Samos, who lived from 310 to 230 B.C. He claimed that the sun and stars were stationary and that the Earth and the other planets followed orbits round the sun. The Earth's circumference was measured with surprising accuracy by Eratosthenes who was a contemporary of Aristarchus and succeeded him as librarian at the great cultural centre established by Alexander the Great in the city of Alexandria, which bears his name.

The man who, more than any other, may be said to have founded the science of astronomy, was also a Greek; Hipparchus, born in 160 B.C. At a time when few accurate measuring instruments were available he calculated the inclination of the earth's axis by studying the recorded times of sunrise and sunset in various places and at different seasons. He produced some of the earliest star-maps and discovered the slow movement of the earth's axis known as the precession of the equinoxes. This movement originates in the equatorial bulge on the Earth's surface which affords a kind of lever for the gravitational pull of the sun and the moon. They are, in effect, trying to bring the Earth's axis upright, but, because of the planet's motion, the result is to make the axis perform a slow circuit of the sky above both poles. This produces a succession of different 'pole stars' over a period of 26,500 years.

Another eminent Greek, associated with Alexandria, was Ptolemy, who lived around 125 to 150 A.D. He made revisions of the star catalogues which had been compiled some two and a half centuries earlier by Hipparchus. He gave his name to the surprisingly resilient and intricate explanation of the movements of the planets, the Ptolemaic System, which survived without serious challenge for 1,500 years until it was finally supplanted by the teaching of Copernicus.

Ptolemy was the last great astronomer of the ancient world. His life marks the concluding chapter of classical science and the onset of the Dark Ages. His famous work, known as the *Almagest*, is one of the few scrolls to survive the fire which destroyed the library of Alexandria. Nothing is more calculated to evoke admiration for these early watchers of the sky than to observe for ourselves how puzzling and complex seem to be the movements which take place in the heavens.

In order to identify the old constellation figures and watch their changing sequence through the year, to follow the motions and phases of the moon, the appearances and disappearances of the Earth's neighbouring planets, no optical aid, such as a telescope, is needed. The naked eye has a wider field of view than any optical device and the observer who has exploited its use fully will be better able, in due course, to appreciate the powers of his telescope.

Tracking the Stars

In clear weather we can watch, each day, the sun passing along its east to west track against a featureless background of blue sky. The brilliance of the sunlight in our atmosphere prevents us from appreciating that it is not just the sun which is taking part in this motion. If we were able to see the stars as well it would be apparent that the sun's stellar back-cloth of constellations is moving with it, though not in complete uniformity. Observing the sky in this way we should discover that the sun's position among the stars is changing. It is following an eastward path among the stars, a track known as the ecliptic, which carries it through the twelve star-groups of the Zodiac. These are often referred to as the Signs of the Zodiac, although this causes some confusion since the astrological 'signs' no longer coincide with the actual constellations. This is a consequence of the changing direction of the earth's axis brought about by the precession of the equinoxes.

At the beginning of spring, thousands of years ago, the sun was moving through the constellation of Aries, the Ram, and its position on March 21st is still designated the First Point of Aries though the actual point now lies in the neighbouring constellation of Pisces, the Fishes, which is to the west of Aries. The sun's motion does not carry it into the stars of the Ram until late in April, 'the princely Ram, glittering in golden wool' as one old poet called it, for Aries was the legendary ram of the Golden Fleece.

This annual circuit of the sky by the sun is, of course, a reflection

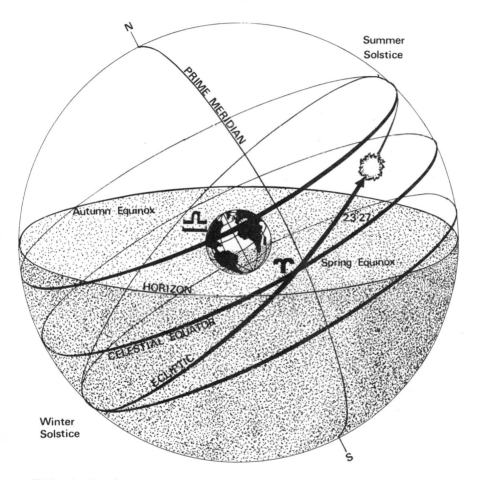

FIG. 1. Earth, Sun and Seasons. The Celestial Equator is the plane of the Earth's equator extended to the sky. The Ecliptic is the plane of the Earth's orbit similarly extended. It is the path which the sun appears to follow during the course of a year and passes through the twelve constellations of the Zodiac.

of the Earth's yearly revolution, and the sun's changing background is produced in exactly the same way as that of a lamp which we might place on a table and walk round. We then see the lamp projected against each of the four walls in turn. Were we unaware of our own movement, as is the case with the Earth's motion, we would naturally assume that the lamp had made a circuit of the four walls.

The movements of the celestial sphere can thus be understood as a consequence of the Earth's two motions. The star-sphere, or that part of it which we can see from any given position on the Earth's surface, makes one complete revolution each day as a result of the Earth's daily rotation, plus a small part, amounting to about one degree each day, of the additional revolution which occurs annually by reason of our passage round the sun. If we align a particular star with some object, such as the top of a spire, or chimney-pot, and note the time by the clock, we shall find that it arrives at that position four minutes earlier each succeeding night and about two hours earlier after the passage of a month. This produces the seasonal changes among the constellations during the course of the year. In the northern sky, however, there are some star-groups such as the Plough or Dipper, part of the larger constellation of Ursa Major (the Great Bear), which never sink below the horizon of observers in northern latitudes. These are the circumpolar stars, while the Pole Star itself appears to remain in the same position throughout the year. In fact, however, since it is not precisely at the north pole of the sky, it does make a very small circuit, though this is not obvious to the unaided eye. If our latitude is about 50 degrees north of the equator, the stars within 50 degrees of the pole will always be above our horizon. Having established the way in which the stars follow their apparent courses across the sky, let us look at some of the other features of the heavens as they can be seen with the naked eye.

Tracking the Moon

The moon is the most conspicuous object visible in the night sky, although to the casual observer its movements and changing phases are sometimes puzzling. The slender crescent of what is popularly called 'the new moon' appears in the western sky after sunset and on succeeding nights, as darkness falls, it will be found farther to the east with a wider crescent, until it eventually appears above the eastern horizon, at sunset, as the full moon. After this, the hour of its appearance above the horizon grows later, with the illuminated area becoming smaller in its waning phase until it is lost in the rays of the morning sun.

The path of the moon round the earth is elliptical, like that of the Earth round the sun. The Earth is situated at one of the two

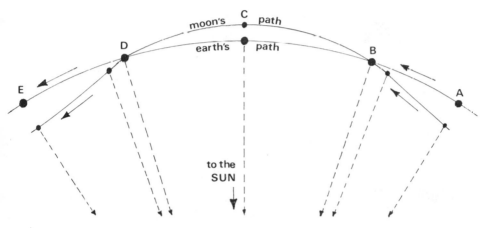

FIG. 2. Earth and Moon: Although the moon's path in relation to the Earth is an ellipse, both bodies are moving round the sun. As a result the moon's track is a curve of constantly-changing radius but always concave to the sun.

foci of the ellipse. At its nearest point, perigee, the moon's distance is 222,000 miles and at the farthest point, apogee, extends to 253,000 miles. Its orbital speed is 0.64 miles per second. The time it takes to complete its circuit of the Earth may be given in two ways. The Synodic month is the interval of approximately 29½ days between one full, or new, moon and the succeeding one, when the sun, the moon and the Earth are all in line. The Sidereal month is shorter, approximately 27¼ days, and, in this instance, a star is used, instead of the sun, in line with the Earth and the moon.

The description of the moon's orbit round the Earth as an ellipse, is a simplification, for both bodies are in orbit round the sun and the moon's track is extended by this motion into a long curve of constantly changing radius but always concave to the sun, as may be seen from the diagram.

As our satellite (the moon) revolves around the Earth, it is also

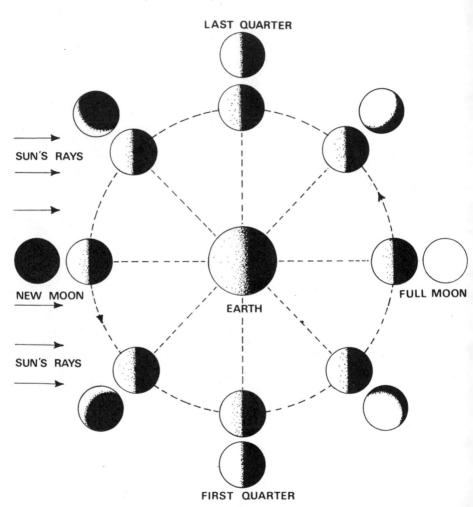

FIG.3. *The Changing Phases of the Moon as it revolves round the Earth each month.*

rotating about its own axis, the two periods of time involved being the same. This means that the moon always turns the same hemisphere towards us, though owing to certain irregularities in its motion we can see a little more than half of its surface over a period of time. Sometimes it is possible to see a little more round one side of the moon. This is termed the moon's libration in longitude. It is caused by the 'spinning' part of the moon's motion getting a little ahead of, or behind, the 'revolving' part, since, although the rotational speed is uniform, the elliptical orbit produces a speed of revolution which varies slightly with the moon's distance from the Earth. There is also libration in latitude which allows us to see a little over the top of, or under, the moon's globe, because its axis is tilted slightly with respect to the plane of the lunar orbit. Altogether, owing to the two librations, we can see almost sixty per cent of the moon.

To be accurate, we never see the moon when it is 'new', since it then lies between the Earth and the sun and the hemisphere turned towards us is in darkness. Only when its eastward motion carries it away from the sun do we begin to see part of the illuminated hemisphere as the familiar, narrow crescent. It is at this stage, too, that we can see the 'earthshine', the reddish glow lighting up the larger segment of the globe, 'the old moon in the new moon's arms' as an old ballad describes it; sunlight which is reflected from the Earth's oceans, clouds and continents to the barren surface of the moon and back again to catch the eye of the evening stroller.

Tracking the Planets

It was with good reason that the ancient world described the five bright planets of which they knew as 'the wandering stars'. These were Mercury, Venus, Mars, Jupiter and Saturn. They were unaware of the other three revolving outside the path of Saturn: Uranus, discovered by Sir William Herschel in 1781, Neptune discovered telescopically by Galle of Berlin in 1846 after its location had been determined mathematically by Adams and Leverrier, and Pluto whose existence was first detected on a photographic plate by Tombaugh, the American astronomer, in 1930.

Although the planetary orbits are all technically elliptical in shape, as Kepler's first law of planetary motion declared, they do not depart greatly from circles. They all lie in nearly the same plane, with the notable exception of Pluto, which follows a path inclined at an angle of 17 degrees to the ecliptic — or the plane of the Earth's orbit. One result of this exceptional behaviour of Pluto is to make it the only one of the planets which moves outside the Zodiac belt of the sky. Pluto's orbit is also eccentric to the extent of bringing it, at times, inside the path of its inner neighbour, Neptune.

Viewed from the Earth the planets fall into two groups; the inner pair, Mercury and Venus, and the outer group consisting of all the others. This is the material factor in deciding when and where we can find them in the sky. Mercury, being nearest to the sun, spends almost all of its time above our horizon during the hours of daylight, overwhelmed by the brightness of the sunlight. It becomes visible in the evening sky only for a few days around the dates of its greatest apparent distance from the sun, the point of eastern elongation. Similarly it can be seen in the dawn sky for a few mornings around the date of its western elongation, rising in the east before the sun. Even so, its greatest angular distance from the sun can never exceed 29 degrees. Neither is every elongation favourable for every latitude. In Britain and other northern latitudes it is best seen as an evening object in spring, when the ecliptic, and therefore Mercury's own path, makes a fairly steep angle with the western horizon and the planet is above the horizon for a longer interval than at other times of the year. In the morning sky it is mostly easily observed in the autumn when its rising precedes that of the sun by the maximum possible time. In the southern hemisphere the positions are reversed.

Venus, though a much larger world than Mercury, is also best seen near its elongations, to the east of the sun in the evening sky and in the eastern morning sky when west of the sun. It also travels farther from the sun with a maximum angular distance of 47 degrees. Both Mercury and Venus exhibit changing, crescent phases resembling those of the moon, though these are not visible to the naked eye. Even before these crescent phases were seen after the invention of the telescope, it was realised that they should be present if the

Copernican theory of the solar system was accurate and the revelation by Galileo that Venus *did* show such phases, was vital evidence in support of Copernicus.

The outer planets, lying on the far side of the sun, may also be unfavourably placed for observation when they are in the daylight sky. Conversely, they are most favourable seen when they are on the

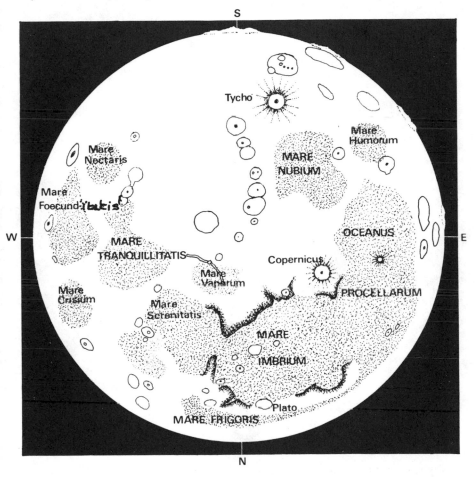

FIG. 4. The main features of the Moon as they appear in a small telescope. The view is the inverted one normally seen with an astronomical eyepiece. Binoculars show north at the top as with the naked eye.

same side of the sun as the Earth. When the Earth lies between an outer planet — such as Jupiter, Mars or Saturn and the Sun — the planet is said to be in opposition. It comes to the meridian at mid-

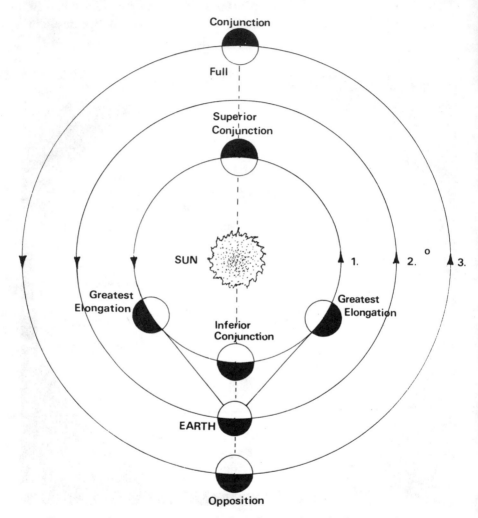

FIG. 5. *The movements of the planets in relation to the Earth and the Sun. 1. the two inner planets, Mercury and Venus. 2. path of the Earth. 3. path of an outer planet such as Mars, Jupiter and Saturn. The inner planets are best seen near elongation and the outer ones near opposition.*

night and is also at its nearest point to the earth. Since, however, the planetary orbits depart from a circular shape there are some oppositions which are more favourable than others and this applies, particularly, to oppositions of Mars. These occur at intervals of 780 days but, because the orbit of Mars is somewhat eccentric, its distance can be as little as 35 million miles when Mars is near perihelion (nearest to the sun) or as great as 62 million miles at aphelion (farthest from the sun). Since Mars is a comparatively small body, the factor of distance becomes an important one for the telescopic observer. The most favourable oppositions of Mars occur in the month of August, although in the northern hemisphere this advantage is somewhat offset by the fact that the planet does not rise so high in the sky as it does in the winter months. This also applies to the other outer planets. An opposition of Jupiter or Saturn in summer means that it remains low in the southern sky, where the denser layers of the atmosphere reduce its brightness.

Although Uranus is, at opposition, theoretically within reach of the naked eye, it lies very close to the limit of visibility, so presenting a comparatively unattractive target for the naked-eye observer and even a telescope of moderate size will reveal very little beyond a small, sea-green disc. Neptune and Pluto, of course, both lie too far away to be seen by the unaided eye and are only of academic interest. Apart from the stars, planets and the moon, the naked-eye observer will find a few other subjects for his attention, which have, moreover, a certain charm due to their unpredictable appearances.

The Aurora

In the Earth's northern latitudes we find the spectacular sky phenomenon named the Aurora Borealis, or 'northern dawn'. This term was first applied to it by the French philosopher, Gassendi, in the seventeenth century. A similar phenomenon in the southern hemisphere is called the Aurora Australis.

The Aurora looks like a luminous glow in the upper layers of the Earth's atmosphere. It is seen most often, and in its most brilliant form, along a belt situated about 25 degrees of latitude from the poles. In the northern hemisphere this takes in the far north of the

American continent and oceans lying off northern Europe, but, since the glow is high in the atmosphere, from about 50 to 200 miles above the ground, it is normally visible over a comparatively wide area.

The Aurora can appear in a variety of forms. It may be no more than a faint, dawnlike glow above the northern horizon in the latitude of Britain or it may brighten into a display of exceptional brilliance, especially near the sun's peak of sunspot activity with which it is closely associated. At such a time it may develop into wide-sweeping arcs of colour; red, green, blue or white. At other times it can produce long rays like the beam of a searchlight or hang like folds of coloured curtains in the sky.

The Aurora is caused by the streams of charged particles which are ejected from the sun's surface in what has come to be called the solar wind. The Earth's magnetic field sweeps them towards the polar regions where they make the tenuous, upper layers of air become luminous like the gas in a sodium vapour street lamp or a neon shop-sign.

At the extreme edge of the Earth's upper atmosphere are the regions known as the Van Allen belts where the charged particles are trapped by our planet's magnetism. They exist at a height of about 600 miles and were the first major discovery produced by America's programme of satellite launchings, which began with the Explorer I satellite in January 1958.

The disturbances on the sun which cause these displays of the aurora may also produce other effects. The upper atmosphere, the layer known as the ionosphere, performs a useful purpose in reflecting radio waves transmitted between widely-separated parts of the earth. The bombardment of this layer under the influence of a solar flare has a damaging effect upon its powers of reflection and, instead of being able to travel round the Earth's curved surface, the radio waves continue their normal course in straight lines and are lost in space or absorbed in the upper atmosphere.

Observers who systematically study the aurora now follow an internationally recognised method of reporting their observations to the appropriate quarter, such as the section of the British Astronomical Association which is concerned with the collection and collation of the information submitted by its members.

The Zodiacal Light

This sky-phenomenon bears a certain resemblance to the aurora in its delicacy and, so far as temperate latitudes are concerned, also in its erratic occurrences. In fact they have common origins, since both are very closely involved with the outflow of particles from the sun in the solar wind. Though seen more frequently in the equatorial regions there are records, as far back as the time of Kepler, reporting its appearance in Europe. Kepler, indeed considered it to be an extension of the sun's atmosphere and, in the circumstances, his conclusion was a remarkably shrewd one.

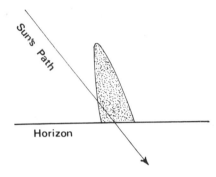

FIG. 6. In north temperate latitudes the Zodiacal light is most likely to be seen in the western sky after sunset in early spring, when the sun's path is steeply inclined to the horizon. In the southern hemisphere autumn is the best time. In tropical latitudes it can be seen throughout the year.

The Zodiacal Light appears like a cone or lens-shaped beam stretching upwards from the eastern sky before dawn, and the western sky after sunset, along the line of the sun's path, the ecliptic or the Zodiac, as its name implies. It is best seen, in the latitude of Britain, from about mid-February to mid-March, or in the morning sky around October. As in the case of Mercury, these dates are reversed for the southern hemisphere.

It may easily be overlooked as a normal aspect of the twilight, but the Zodiacal cone is distinguishable by a more concentrated beam, narrower than the extent of the ordinary twilight illumination and reaching higher into the darker region of the sky. At those times of the year when it is most readily seen, the cone makes an

31

angle of about 60 degrees with the horizon. On other occasions a much smaller angle makes it difficult to detect. Even more delicate is the so-called counter-glow, or gegenschein, a patchy and ill-defined glow sometimes occurring in the sky diametrically opposite to the sun.

When the Zodiacal Light is examined with the spectroscope it proves to be the reflection of sunlight from small particles. Just what these particles consist of is so far undetermined. Very small particles would probably spiral inwards towards the sun, though at times near the sunspot maximum, when the solar wind is active, the tendency would be for the particles to be pushed outwards. There may be somewhat larger particles of matter involved, with diameters of a few centimetres, but whether this material is left over from the time when the planets were formed or whether it is related to comets and meteors, remains uncertain. Whatever its origin it must be conceded that even the debris of the solar system is still capable of producing effects both mysterious and spectacular.

Comets and Meteors

Most phenomena of the night sky are matters of routine. The westward seasonal march of the constellations, the changing phases of the moon, the movements of the planets and the eclipses of the sun and moon are all events which follow a well-defined sequence. Comets and meteors belong, to a very large extent, to the realm of the unpredictable. It is true that some comets, like the well-known one named after the astronomer Edmond Halley, follow orbits round the sun which are understood and their return can be forecast, but most follow orbits of very elongated ellipses which have not yet been accurately computed, or they move along paths which are open curves, such as the parabola and hyperbola. In the latter cases the comet makes one brief appearance in the neighbourhood of the sun and then vanishes for ever into the depths of space from which it briefly emerged.

Although a comet may develop a spectacular tail, stretching behind the head, or coma, for a distance of some hundreds of millions of miles, many never become more than faint, misty spots of light in the lens of a telescope. When the comet nears the sun, its perihelion, the tail begins to develop as the frozen gases and solid particles of the head are affected by the heat of the solar rays and the pressure of the solar wind. By this process the comet loses part of its material. It may divide into two or more parts as it passes near the sun or one of the major planets. This happened in 1947 when a comet came within 10 million miles of the sun and was broken into two pieces. Eventually even the largest comet must disintegrate leaving behind it a swarm of particles which may at some time rush into the Earth's atmosphere to produce a shower of meteors.

Amateur astronomers have often discovered new comets and here a large pair of binoculars is frequently found useful, the essential requirement being a large light-grasp and wide field of view rather than high magnification. The most favourable area in which to hunt for comets is in the western sky after sunset or in the eastern sky before dawn, but care must be taken to avoid identifying any vaguely-defined misty spot in the sky as a new comet. It was to forestall this eventuality that the French astronomer, Charles Messier, in the eighteenth century, compiled his list of nebulous objects in the expectation that this would help him to identify the real comets when they came his way. By an odd stroke of irony it has been Messier's fate to be remembered not for his comet discoveries but for his list of non-comets.

A comet, of course, will change its position among the stars after a short interval and from night to night the difference will be very noticeable. The remote nebulae and star-clusters in Messier's list, like all other objects of their kind, show no such movement. A comet may be moving at a speed of about 200 miles per second when it is near the sun, but its motion is nothing like that of a meteor flashing across the sky. The latter is a small fragment of stone or metal, perhaps no larger than a grain of sand, but the luminous trail it traces across the sky is very close to us as it passes through the atmosphere to burn up. The comet, though travelling at a much greater speed, and of vastly greater size, is seen at a distance of many millions of miles.

A few meteors may be seen at any time of the year but there are some dates when they may occur in greater numbers or 'showers'.

The Perseid meteors of mid-August and the Leonids of mid-November are two of the best-known examples.

Although these meteor showers, which number about twenty through the year, are identified by the names of various constellations, it should be understood that this in no way implies the existence of any physical connection with the stars. The reason is that the meteors appear to come from that particular part of the sky, such as Leo and Perseus in the two instances already mentioned.

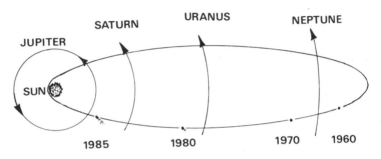

FIG. 7. The path of Halley's Comet. Its last appearance in the night sky occurred in 1910 and its passage was followed by astronomers for almost two years until it had once more moved beyond the orbit of Jupiter.

This is known as the radiant point of the meteors. Other radiants lie in the direction of Lyra, Capricornus, Orion, Taurus and Gemini.

Meteor observation is essentially an activity for the naked-eye observer although with improvements in lens manufacture and greater sensitivity of film emulsions, photography has been employed in recent years. Even the common 35mm. camera, which is normally fitted with a fast lens, has been used with notable success. Sometimes the technique is simply to point the camera towards a meteor radiant, in the case of an expected shower, leave the shutter open and hope for the best. One of the hazards of the process is the probable formation of dew on the camera lens during cold nights.

In the case of visual observation, it is essential to record, as accurately as possible, the direction of the meteor's flightpath and its duration. This is transferred to a chart of the sky and a series of such observations will enable the radiant point to be determined by extending the lines of flight backwards. The time, estimated brightness on the scale of star-magnitudes and any information as to the appearance of the meteor — whether a quick flash or a long, luminous trail — are other points to be noted.

The number of known radiant points runs to several hundreds and every month of the year produces at least one shower, though these are not equally prolific.

In this connection the use of the word 'shower' is to some degree misleading. We may watch the radiant for an hour and see only three or four meteors but there have been occasions when the 'shower' has really justified its name. Such was the case with the Leonid meteors of November 1833 and 1866. It seemed that this display could be relied upon to provide something really spectacular every 33 years and hopes were raised that the event would be repeated in the year 1899. In the event the expected torrent of falling stars proved to be only a mere trickle and it was not until 1966 that the number of meteors again rose significantly, with many thousands being seen each hour in parts of the United States. In Britain, however, the peak of the shower was missed by the expectant watchers for it took place during the hours of daylight.

The enthusiast should also bear in mind that meteors are more frequently encountered between the hours of midnight and dawn than during the evening. After midnight the Earth's rotation is carrying us in the same direction as its orbital revolution and we then encounter the meteors 'head-on' instead of being overtaken by them as happens in the evening hours. In consequence the morning meteors rush into the atmosphere at greater speed and they tend to be brighter as well as more numerous than those we are likely to see at an earlier, and probably more convenient, time. Large meteorites, substantial pieces of stone or metal which survive their passage through the air to reach the ground, are, fortunately, rare. The most notable example of recent years was the Barwell meteorite of December 24th 1965 which broke up after crossing the English Midlands and scattered its fragments over the Leicestershire countryside.

2 The Seeing Eye

It is remarkable that our view of the world around us should be conveyed to our brain via so small an aperture as that of the pupil of the human eye. Its size is variable, like the iris diaphragm of a photographic lens, though this opening and closing occurs involuntarily, stimulated by the light which the eye encounters. At maximum opening, in darkness or a very dim light, it may reach about one third of an inch in diameter though the precise degree of adaptability naturally varies with individuals and declines with advancing age. There is some difference in sensitivity in parts of the retina which is the inner, rear surface of the eye. The retina is composed of nerve-endings which are connected to the brain by the optic nerve at which point there is a 'blind spot' where the retina is insensitive to the stimulus of light-rays. The nerve-endings are of two kinds, named the rods and cones. The rods are the nerve-endings which are most sensitive to faint light-sources. This has important implications for the astronomical observer who is constantly concerned with remote and faintly luminous objects. Most of the rods are present in the outer part of the retina while the colour-sensitive cones occupy the central area.

The eye is not equally sensitive to all colours. It is more easily impressed by blue light than red and if we look at a red rose in a dim light it will appear black, although its outline may be clearly perceptible. This is sometimes called the *Purkinje Effect* after a 19th century Czech physiologist of that name. This characteristic of the human eye also has its significance for astronomical observers, especially those who make a particular study of variable stars and are frequently concerned with making visual estimates of a star's changing light, one field of astronomy to which amateurs are continuing to make effective contributions.

The sky contains many subjects for the observer who wishes

to test his powers of eyesight, with or without a telescope or pair of binoculars. The Pleiades star-cluster is a good example. These stars are sometimes called the Seven Sisters though many people can see only six and there are a variety of legends in the folklore of astronomy accounting for the 'lost' member of the group. Nevertheless, good eyes will see seven stars without much difficulty and exceptional sight is capable of detecting ten. For the purpose of picking up faint stars it is useful to turn the eye slightly aside from the direction of the object sought, since this brings the light-sensitive outer part of the retina, the rods, into more effective use. Another test is provided by the area enclosed in the bowl of the Dipper in the northern sky. Good eyesight should be able to see at least six stars within it but keen sight may count several more. The result will obviously be affected to some extent by the altitude of the group, the conditions of the atmosphere and the presence or absence of moonlight or other extraneous light-source which will inevitably reduce the eye's sensitivity. Jupiter's satellites have often been sought by naked-eye observers but the evidence for their visibility seems somewhat unconvincing, though it is true that once an object has been seen with adequate optical aid it can often be seen without it or with less assistance.

Apart from its light-gathering function the eye is also concerned with the resolution of detail, the ability to separate points. A newspaper, for example, held at a distance of several yards may show the headlines distinctly because the type is large, whereas the small print is seen only as a blurred, greyish patch. To distinguish the words and letters in small type we must get closer to it or use some kind of optical magnifier. Here also the sky provides some interesting tests for the eye's resolving ability in the form of double stars. In the handle of the Dipper (or the shaft of the Plough) the second star from the end is Mizar. It has a smaller companion star which was once said to have been used by the Arabs as a test of eyesight. If this was so, the test seems to have been a relatively undemanding one, for the small star, Alcor, can be seen quite readily by people of normal visual acuity. It is also sufficiently distant from Mizar to present no problem of separation.

A more difficult test will be provided by a much fainter and closer pair of stars in the little constellation of Lyra, the Lyre, which rises high into the south during the summer months. The star is called Epsilon Lyrae and to see the two components clearly and distinctly as a pair requires good sight. The same star is of great attraction to the telescopic observer, for each of the two is divided into a close double star, while between them there are two more extremely faint ones which will test the eye and also the telescope's light-grasping powers unless it happens to be a fairly large one. The separation of double stars is measured in seconds of arc. The two stars in the naked-eye double are 208 seconds apart but each of the close pairs has a separation of only a little over 2 seconds of arc.

More Power to the Eye

The first purpose of an optical instrument such as a telescope or a pair of binoculars (essentially a pair of small telescopes placed parallel to each other), is to collect more light than the eye is capable of collecting alone and to make it possible for this greater quantity of light to be directed into the pupil of the eye with little waste. Of course it often happens that instead of the human retina receiving the light, it is directed, instead, to that artificial kind of retina which we call a photographic plate or film, where it makes a permanent impression.

The light-collecting task is performed by the object-glass of a refracting telescope, or by the mirror in the case of the reflector. In both instances this is called the objective, or the image-forming element of the telescope. It is obvious that the amount of light collected from a given source, such as the sun, moon, planet or star, will depend upon the size of the objective, so, other things being equal, the bigger the object-glass or mirror, the better. But this is merely a theoretical deduction for although professional astronomers now have very large telescopes, and they would like still bigger ones, other things are by no means equal and in each individual case a compromise has to be arrived at. The largest mirror which is operating in a telescope is the 236-inch diameter objective of the great Russian telescope. The largest object glass is only a fraction of this size, the 40-inch lens of the Yerkes Observatory telescope in the United States. A larger lens could doubtless be made but the Yerkes lens, supported along its circumference, already sags measureably under its own weight when pointed to the sky.

So far as light-gathering power is concerned, the area of the object-glass or mirror is the determining factor, so that when we are comparing two objectives we can say that the light-grasp varies as the square of the diameter. A four-inch lens may have only twice the diameter of the smaller two-inch lens but their areas, and hence their light-grasp, will be in the proportion of sixteen to four or four to one. Such differences as may occur between the lens and a mirror of equal size will be considered later. Comparisons between the refractor and reflector have aroused a great deal of controversy in the past and the issue is by no means settled yet. In any case the matter is one which involves a number of varied considerations, not least of which is the cost.

The resolving power of the lens or mirror also increases with the size. A large telescope will show more fine details on the moon, or on a planet such as Jupiter, Mars or Saturn, than a small one, always assuming that we are considering instruments of comparable optical quality and not a bad large one with a good small one.

How Big?

It has already been made clear that the largest possible objective is desirable from the point of view of light-grasp and theoretical resolution of detail. There are, however, other relevant points to be considered. Cost is one of them. With increase of size, particularly in the case of the object-glass, which has four, or even six surfaces to be ground and polished, the cost will increase very rapidly. So will the problems of mounting the telescope effectively as well as the difficulty of moving it, if intended to be portable, or housing it in some kind of shelter or observatory if this is necessary. Before considering the matter in detail it may be useful to look at the possibilities offered by the very simplest form of telescope, like that used by Galileo and others after his time. Although there are unlikely to be any important new discoveries to be made with very small telescopes now, nevertheless, there is much to be said for beginning one's observing career with modest means, first with the naked-eye, then with a pair of binoculars or a small telescope before graduating to a larger instrument. The author of the century-old classic of amateur astronomy, *Celestial Objects for Common Telescopes,* the Rev. T.W. Webb, advised his readers never to disregard even diminutive instruments. They will, he said, show something never seen without them.

The very small telescope may not possess a high magnifying power, but we need not worry overmuch about that. A lot can be seen with a low power, perhaps 10x or 20x. Some of the larger formations on the moon will be visible. If used as a projector and NEVER for direct observation even with a dark screen, some of the larger sunspots will be seen. Venus will show its crescent phase, Jupiter a perceptible disc and its large satellites. We may just glimpse the form of Saturn's rings when they are fairly wide open, but, above all, the small telescope is capable of a remarkably impressive performance on the stars; by no means as a divider of double stars (except a few very widely-separated pairs), but as what has come to be called a 'rich-field' telescope.

The R.F.T.

Even in a small telescope it is desirable that we should obtain a lens of reasonable optical quality. It should, if possible, be about two inches in diameter and be an achromatic lens of two components. A single lens will be afflicted by problems of unwanted colour, an objection which made Sir Isaac Newton turn to the mirror form of telescope which is happily free from what is termed chromatic aberration. Early attempts to overcome this problem by making the lenses of very long focal length involved some very ungainly arrangements, such as mounting the lens at the top of a high pole while the observer attempted to align the eyepiece with it from the ground, many feet below.

It is still possible to obtain small lenses of excellent quality from ex-government sources, as well as some eyepieces of fine performance. Age is no barrier to good definition. The writer's three-inch refractor was made by the firm of Dollond in London more than a century ago and still works admirably. Neither need we worry

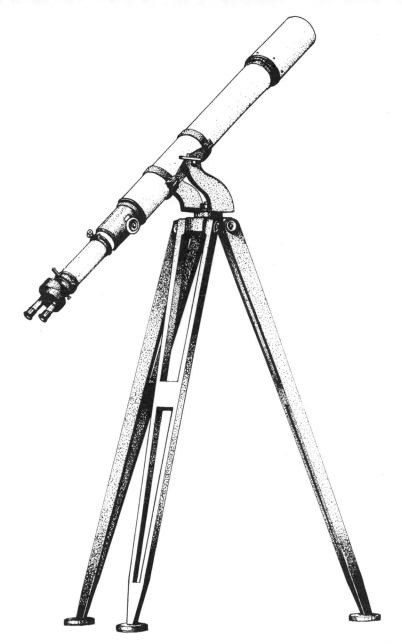

FIG. 8. *A small astronomical refracting telescope on altazimuth tripod mount. This instrument has a three-element photo-visual object-glass of 3½-inches aperture by the famous old maker, Thomas Cooke and Sons. It is shown with a Zeiss binocular eyepiece attachment.*

41

unduly over a few scratches, for these will only intercept a very small amount of light.

The two-inch lens may be mounted in a tube of strong cardboard such as a postal tube and the eyepiece in a smaller one to slide in the other end for focussing the image sharply. The simple telescope will prove highly effective in providing views of star clusters, the Milky Way, the Orion Nebula and the Great Andromeda galaxy. It will, in fact, show more stars at once than larger telescopes because of its wider field and there are many observers with instruments of considerable size at their disposal who find satisfaction in the use of a small 'RFT' or a pair of binoculars of the 7 x 50 specification which offer a comparable performance.

My own first steps in astronomical observation were, in fact, taken with a somewhat less effective instrument, an old pair of Galilean field-glasses of First World War vintage, the forerunner, as it turned out, of many optical ventures. This has probably been, in some measure, the experience of many enthusiasts, but I cannot whole-heartedly accept the assertion that no properly constituted amateur astronomer will be satisfied until he owns the biggest telescope that he has time to use, space to house and money to buy. Here, indeed, we come to the crux of the matter so far as size is concerned.

It is regrettable, but true, that optical instruments, being in the main composed of fairly expensive material — such as high-quality optical glass — and less amenable than some products to large-scale processes of manufacture, have been subject to the same inflationary pressures as most other commodities. In 1955 I was fortunate enough to be able to buy a fine 5-inch refracting telescope made by the famous old English firm of Thomas Cooke and Sons. It had a three-element object-glass of the type known as photo-visual, on a heavy mahogany tripod with altazimuth mount and slow motion handles. The cost at that time, though admittedly at secondhand, was £150 or about $280. Now that sum will only buy one of the cheaper pairs of binoculars in the West German Zeiss range.

My experience with this fine telescope quickly taught me the necessity of compromise in the matter of size. A five-inch object-glass has a substantial light-grasp for most purposes and will give outstanding views of the moon and planets. Today, unfortunately,

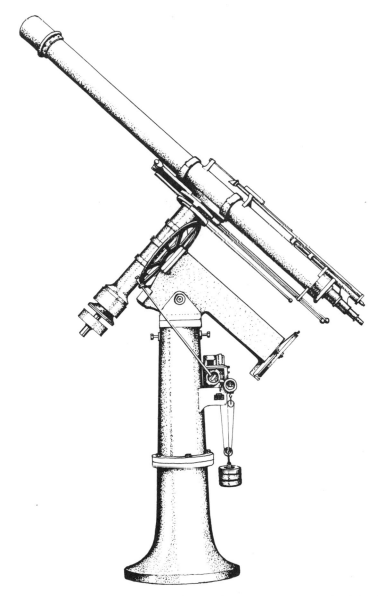

FIG. 9. A six-inch refracting telescope on equatorial pillar mounting with slow motion controls and driving-clock. An instrument of this size requires an observatory, but for most amateurs a fixed observing site has serious disadvantages when objects become lost among trees or chimney-pots.

it will cost several hundred pounds to buy, if adequately mounted. Even if it can be afforded there is still the question of convenience in use and storing it when not in use.

The Cooke photovisual object-glasses were probably unsurpassed in quality by any in the world, before or since, but they were made in relatively long focal lengths. The 5-inch was mounted in a tube of solid brass which had an overall length of more than seven feet and was proportionately heavy. To remove this from its case, carry it outside on a dark night while evading a variety of projecting corners and obstacles was by no means a light task. Raising it to a height and position which would allow the close-fitting trunnion axles to drop into their precisely engineered slots at the summit of the six-foot tripod, proved both nerve-racking and physically exhausting. Even then the eye-end of the long tube came so low that it was practically impossible to see anything without the use of a star-diagonal when the tube was directed more than a few degrees above the horizontal. A refractor of this size requires a permanent mounting on a substantial pillar beneath some kind of observatory, or at least a run-off shed.

That will save a good deal of physical effort and inconvenience. Dismounting a seven-foot brass tube and carrying it indoors with frozen, gloved fingers, is even more troublesome and worrying than bringing it out.

The fixed mounting, however, in spite of its undoubted convenience, can bring some problems of its own, though much will depend upon the particular site. An unobstructed view of the southern sky is the ideal, but many observers will find it by no means easy to choose a suitable spot in any direction, what with trees, roofs, chimney-pots and street-lighting obstructing the fixed point of observation.

Not every five-inch telescope is a refractor in a seven-foot brass tube and seemingly a ton in weight. Some years after disposing of the Cooke instrument I located (in the small advertisement columns of the press) the exact contrast. This telescope had a lens of six inches diameter at the front. It was mounted in a light aluminium alloy tube no more than two feet in length and even mounted on its tripod could be carried in and out of the house with ease. In

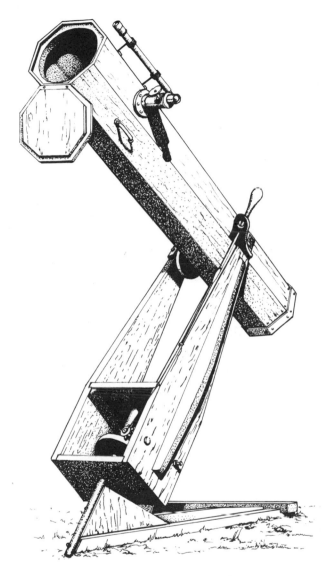

FIG. 10. A Newtonian reflecting telescope of medium size on equatorial mounting of fork type. The octagonal wooden tube is well-suited to its function. The mount is fitted with a slow motion control in right ascension.

spite of its large front lens it was quite a different telescope from the refractor and had its image-forming element, a concave mirror, at the lower end of the tube. This was my first introduction to that compact and highly-effective modern telescope, the Maksutov, of which more will be said later.

In all the literature of astronomy there is surely no more entertaining book than the one called *Of Telescopes* by Dr. William Kitchiner which was published in London by George Whitaker of Ave Maria Lane in the year 1825. Though obviously out of date in certain respects, it can still be read with great profit and interest by any telescope enthusiast. Its author had a keen sense of humour and that he was a man of wide tastes is clear from the title-page, which lists among his published works such titles as *The Cook's Oracle, The Housekeeper's Ledger, The Art of Invigorating and Prolonging Life* and *The Pleasures of Making a Will*. On the subject of large telescopes, however, Kitchiner had firm opinions. They are, he said, of no more use than the huge spectacles which opticians hang outside their shops. 'What good', he asks, 'can a great deal of bad light do'?

Copies of the old doctor's book are now difficult to find and I was especially fortunate to obtain the copy which once belonged to W.F. Denning. Even though it has moved into the category of 'collectors items' it can still be found in astronomical libraries.

3 Choosing a Telescope

Although producing the accurate optical surfaces which are required for good performance in a telescope can involve some highly refined and sophisticated techniques, the optical principles involved are relatively simple. Parallel columns of light from an object in space fall upon the curved surface of either a lens or a mirror. In the former case they are bent, or refracted from their parallel course to a point called the focus. The mirror, too, has its focal point, to which the light-rays converge after being reflected back by the coating of silver or aluminium on the mirror's surface. At the focal point an image is formed which is very small and bright. The eyepiece is only a form of magnifying-glass, though it may take a variety of designs with anything from one to six component lenses. This magnifies the image and converts the light into a concentrated beam of parallel rays which passes into the eye.

The refractor used by Galileo in the seventeenth century represents the telescope in its most elemental form, but the modern large refractor still acts in the same way. The optical parts are now bigger and more complex in their design, but, though Galileo himself would undoubtedly be impressed by the advances made, he would find it surprisingly like his own telescope in general outline and manner of use.

The most common design of reflecting telescope is still that of the Newtonian, although Newton was not the first to envisage the use of mirrors for the purpose. The Scottish mathematician, James Gregory, had in 1663 drawn up a design for a shorter form of reflector which still bears his name, the Gregorian, now more or less obsolete except as a museum-piece for collectors.

Both of the main telescope forms have now been in use for over three hundred years and there is still no definitive answer to the oft-repeated question as to which is the better. There are as many answers as there are individuals seeking the best telescope for their own purposes. We can, however, easily set down the plus and minus factors for each type, although the choice is now somewhat wider than just between the refractor and the Newtonian reflector.

The colour faults of the lens cannot be completely eliminated even in the costly, three-element photo-visual type of objective, but in the sizes normally used by amateur astronomers, with diameters from three-to six inches, the problem is not a serious one. Moreover, anyone who uses a refractor fairly regularly becomes so accustomed to any slight, residual colour error that it passes without notice.

Refractors are normally made with focal lengths around fifteen times their aperture. Long focal lengths tend to reduce the colour problem. This, in turn, requires a long tube in sizes over three or four inches, with the difficulties of moving, mounting and housing, already mentioned. It has the advantage of a closed tube along which the light rays must pass only once, and this tends to produce a steady image. The long focal length also gives some advantages in the matter of eyepieces, since these too can have a moderate focal length while still giving a reasonable degree of magnification. They can be of a simple type and therefore less expensive, but it is false economy to look always for the cheapest eyepiece. The strength of the chain is that of its weakest link and the quality of performance in a telescope is that of its weakest optical element.

When we come to consider the all-important factor of cost, there can be no doubt where the superiority lies. A Newtonian reflector of eight or ten inches aperture will cost less than a four or five-inch refractor. If we think of the matter in terms of power per pound or dollar of expenditure, the reflector must be the acknowledged winner. The chief reason for this lies in the fact that the mirror can be made of much cheaper material, such as plate glass, pyrex or some other forms of low-expansion glass. Furthermore, only one surface must be worked, instead of the four or six surfaces of the expensive, optical glass which go to make the object-lens.

Likewise, the mirror can be made of much shorter focal length without running the risk of colour troubles. Though the refractor with a two-element objective is often called the achromatic telescope, its name is, strictly-speaking, a misnomer. The truly achromatic telescope is the reflector. Light of all colours is brought to the same focal point by reflection from a mirror. The lens is unable to achieve this and the rays which are not brought to a focus in the refractor inevitably affect the quality of the image. Where colour estimates are involved, as frequently occurs in observations of the moon, planets and stars, the mirror is, unquestionably, the final arbiter.

If the focal length of the mirror is six times the aperture, it is said to work at a focal ratio of f/6. This, or f/8 are common, although there are certain advantages to be derived from the use of a longer focal length, with 'f' ratios of ten or twelve, but a moderate, f/8 ratio has much to commend it and a mirror of 8 or 10 inches diameter can be accommodated in a tube of modest overall size with the added convenience in mounting or moving.

In use, the refractor involves some inconvenience on account of the eyepiece being at the lower end of the tube and the only means of overcoming this lies in the use of the device known as a star-diagonal, which bends the light through ninety degrees and provides a comfortable, downward angle of view. There is some loss of light due to reflection in the prism and there is also a reversal, right and left, of the image, which can be annoying to anyone unfamiliar with it.

Any telescope or pair of binoculars will only produce its best performance when the optical elements are correctly aligned, or collimated. In the case of binoculars the adjustment of the four prisms is critical and it is often possible to find an error in a new pair. If they are used in this condition the result can be eyestrain since the two images are not brought into coincidence. Correcting such an error is strictly an operation for the skilled optical worker. The case with the single tube of the telescope is, fortunately, an easier matter.

FIG. 11. A Newtonian reflector of 18-inches aperture formerly used by the writer. It was made by J.H. Hindle. The focal length is 7½ ft. and the skeleton tube is shown on its original heavy steel mount of altazimuth design. The telescope was later removed to the observatory at Keele University, Staffordshire and mounted equatorially.

The refractor should present few difficulties in this respect. The object-glass can be seated firmly in its cell and, if treated with the normal care due to any good piece of equipment, it should remain in correct adjustment. Refractors should, however, have pairs of adjusting screws which operate on the push-pull principle fitted to the cell, though they may seldom need touching. With small refractors these are often omitted and an error of collimation, should it occur, is a more difficult problem. It will probably never arise, however. The collimation of my old Dollond is perfect after more than a century and it has no collimating screws fitted to the cell of the objective.

The observing position with a Newtonian reflector offers vastly greater comfort, and therefore almost certainly more effective use of the eye, than the refractor. If mounted on an altazimuth tripod, or pillar, the eyepiece in its draw-tube remains in a horizontal position at all altitudes. If the tube is very long this can bring the eyepiece to an inconvenient height when observing an object at a high altitude where it is normally best seen. The largest reflector of which I have personal experience was an 18-inch Newtonian made by J.H. Hindle of Blackburn. It is now at the observatory of Keele University in Staffordshire, mounted alongside a 12-inch refractor which came from the University observatory at Oxford.

The 18-inch reflector, when in my possession, was in a skeleton steel tube and moved on a massive altazimuth mount. The open tube was about eight feet long and, when elevated to near the zenith, necessitated the use of a step-ladder. It managed to avoid the turbulence often produced by air-currents in a closed, cylindrical tube, but had the occasional disadvantage of the mirror being exposed to the formation of dew. There was also the annoyance and frustration which almost invariably accompanies a fixed observing site, when the moon or a planet became obscured by tree branches. Large apertures have the ability to collect more light than small ones and their theoretical powers of resolution are also greater, but these advantages are not always obvious since the large mirror, intercepting a larger column of light, is also more liable to atmospheric troubles which produce unsteady, flaring images. This is unavoidable but decidely less obstrusive with small or medium apertures. Only in the most favourable circumstances is it possible to exploit the powers of the large telescope fully. Sir William Herschel soon discovered this and often preferred to use his small 6½-inch telescope instead of the 48-inch giant in his garden at Slough. The mirror of the latter can still be seen in the Science Museum at South Kensington, London.

Herschel began his memorable career as an amateur in the field of astronomy. Professionally he was a musician and he found it no easy task to construct his first effective telescope since the mirrors of his time were made of an alloy known as speculum metal, a mixture of copper and tin which has been described as 'a peculiarly mean alloy to cast and work'. The only method of testing the curve of these metal mirrors was by trial and error and the emphasis must often have been on error, for Herschel is said to have made more than a hundred fruitless attempts before he was successful.

There was, however, still a problem for the telescope-maker of Herschel's day to keep the mirror in operational condition. The surface easily became dull and tarnished with exposure to the air and moisture. It could be restored only by re-polishing which altered the delicate 'figure' of the surface. In view of the many problems which then beset the telescope worker it is surprising that so much was accomplished.

The modern glass mirror is not exempt from the problem of the tarnished surface, but the aluminium coating which is now usually applied is much more durable than the silver which it replaced. With exposure to the air it forms an oxide which is both durable and transparent, whereas a silver coating eventually turns black. Surprisingly good results have been obtained even with a mirror surface which looks quite badly affected, though the reflective power is inevitably reduced and the coating of a telescope mirror is something which repays careful treatment.

The grinding and polishing of a telescope mirror is a task which many amateurs have found both challenging and engrossing, though depositing the aluminium coating requires the use of a special vacuum chamber. Anyone who feels unwilling to make the mirror can buy the finished article from a number of sources though it is obviously desirable to obtain some assurance that the 'figure' is good enough to produce sharp definition of a planet when used with a suitable eyepiece and flat secondary mirror, under moderate magnification. More will be said about this in the chapter dealing with the application of the telescope. There is certainly some saving of cost if the observer is able to mount the mirror himself, and here should be added a recommendation for the square wooden tube. This tends to reduce the possibility of dew forming on the flat mirror, which is susceptible since it is nearer to the open end of the tube; it also loses heat more quickly than the main mirror, partly for the same reason and partly owing to its smaller mass.* Some small and medium-sized reflectors are fitted with a prism instead of an elliptical flat mirror, but such prisms of the requisite quality are not always readily available and, when found, they usually cost more than the equivalent flat.

When all relevant factors are taken into account the observer's choice of telescope must depend very much upon what he wants from it. For occasional use in observing the moon and planets, or for solar work, the small refractor has much to commend it. It is relatively trouble-free and portable. It is less sensitive to atmospheric turbulence than the mirror, but, if the amateur wants the maximum power for his money and is not averse to giving his instrument a little care and attention, there is undoubtedly much to be said for a reflector of 6 to 10 inches aperture. There are, however, other telescope designs which deserve consideration besides the basic forms of refractor and the Newtonian reflector.

The Compact Telescopes

There are several telescope designs which have the advantage of folding back the light-path as it passes through the tube and enabling a relatively long focal length to be accommodated in a short space. This is done in a pair of prismatic binoculars with a resultant saving of size and weight, though always at the cost of a slight loss of light. This latter consequence is a small price to pay for the added convenience of compactness.

The earliest design for a reflecting telescope, the Gregorian, was

*A square wooden tube is also less subject to internal currents of air and it also allows the builder to attach the eyepiece mount more easily.

of this kind. Like the Newtonian mirror, the Gregorian uses a paraboloid, the form of curve which is capable of concentrating a parallel beam of light at one point, the focus. A spherical mirror will not do this, the centre rays being brought to a focus at a point farther from the mirror surface than those which are reflected from the edges. This is corrected by deepening the centre portion of the

FIG. 12. The small Newtonian reflector of short focus performs well as a rich-field telescope, giving spectacular views of the Milky Way and star-clusters. It uses an eyepiece of low power and wide field. The compact size makes it very portable with apertures to 6 inches.

mirror very slightly, thus eliminating the fault known as spherical aberration.

The eyepiece of the Gregorian telescope is at the lower end of the tube as with the refractor, just behind the main mirror. The light is returned to the eyepiece by a concave, ellipsoidal secondary mirror near the upper end of the tube, passing through a hole in the main mirror into the eyepiece. The image is upright, unlike the inverted image of the normal refractor and Newtonian. The focussing is done by moving the small secondary mirror while the eyepiece is fixed.

The Gregorian design achieved some degree of popularity in the eighteenth century though mostly in fairly small sizes. It was superseded by a telescope of similar appearance, the Cassegrain.

The Cassegrain Telescope

The year 1672, which marked the election of Sir Isaac Newton to a Fellowship of the Royal Society and his presentation to the Society of his own new form of telescope, also saw the arrival of another design, the Cassegrain, which, like the Newtonian, has survived until the present day. Some of the very largest telescopes are of this type, including, rather ironically, the largest in Britain which commemorates the name of Newton at the Royal Observatory.

The Cassegrain was first described in a French periodical and takes its name, like the Gregorian and the Newtonian, from its designer. Shorter even that the Gregorian, it uses a paraboloid for the primary mirror and a hyperboloid shape for the small convex secondary element. The main mirror has a central hole for the light to reach the eyepiece, as with the Gregorian, and both of these forms share with the refractor the drawback of an inconvenient position for the eye when pointed to an object at a high altitude.

The Cassegrain produces a long effective focal length in a short tube and is essentially an instrument for relatively high-power observation of the planets, where its small field of view is no great disadvantage. Amateur telescope makers have tended to avoid the Cassegrain form, partly because the primary paraboloid itself requires very accurate figuring, since it usually works at a focal ratio between

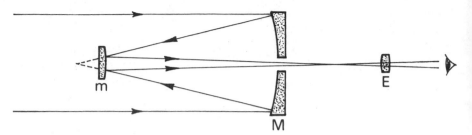

FIG. 13. The Cassegrain form of reflecting telescope has become popular in recent years though its original form has undergone some modifications. M is the main mirror; m is the small secondary mirror and E is the eyepiece. The mounting and observing position are similar to those of the refractor.

f/4 and f/7, and the secondary mirror presents an even greater problem with its hyperboloidal curve.

There are many who have doubted whether it was really worth the effort to make a telescope which has sometimes been said to combine the worst features of both refractors and reflectors. The large observatory telescope is another matter, for there the Cassegrain design has undoubtedly been successful.

The Catadioptric Telescope

As we have already seen, the reflecting forms of telescope described so far all make use of a paraboloidal mirror as the image-forming element in order to avoid the spherical aberration which would otherwise spoil the performance. There are, however, other ways of dealing with this defect, apart from deepening the curve of the spherical mirror into a paraboloid. One such method was suggested by the Russian designer, Maksutov, in 1944.

This makes use of a lens which intercepts the light before it reaches the mirror. The lens has a steeply curved surface which is concave at the front and has an error (positive spherical abberration) which corrects the opposing error (negative spherical aberration) of the mirror. This lens is known as the correcting plate. It is very thick, and since the light passes through it, the glass must be of high optical quality which makes the lens a fairly costly item.

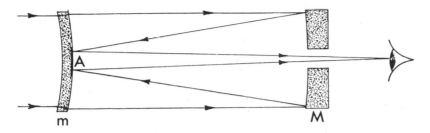

FIG. 14. The Maksutov is a modified form of the Cassegrain design. M is the main mirror of spherical curve; m is the meniscus corrector lens which carries the secondary mirror, A, in the form of an aluminised spot on its inner surface. The Maksutov is a very compact design and capable of a superb optical performance.

The curves of both lens and mirror are spherical which tends to make their production somewhat easier although the mirror has a very short focal length and must be very accurately figured. It is perforated in the centre, as with the Gregorian and Cassegrain, but the secondary mirror is often no more than a small central spot on the inner surface of the correcting plate which is coated with aluminium like the main mirror.

The Maksutov-Cassegrain, as this form of telescope is called, has become increasingly popular and anyone who has used one will probably think that this is deservedly so. Unlike the refractor lens the Maksutov correcting plate (or shell, as it is sometimes called) has negligible colour error, while the compact form makes it easy to mount in small sizes, up to six or eight inches for example. The optical parts are firmly fixed in their cells so that collimation tends to be less of a problem than it is with the Newtonian and the tube is completely closed so that turbulent air currents cannot affect its performance. The Maksutov is a delightful instrument to use on terrestrial subjects as well as for astronomy but the initial cost is substantially higher than for a Newtonian of comparable, or, indeed, greater aperture. In the country of its origin, the Soviet Union, it is perhaps not surprising that the Maksutov design has become highly popular, and it is gaining a growing number of adherents in Britain, in Europe and the United States. For portability it may be

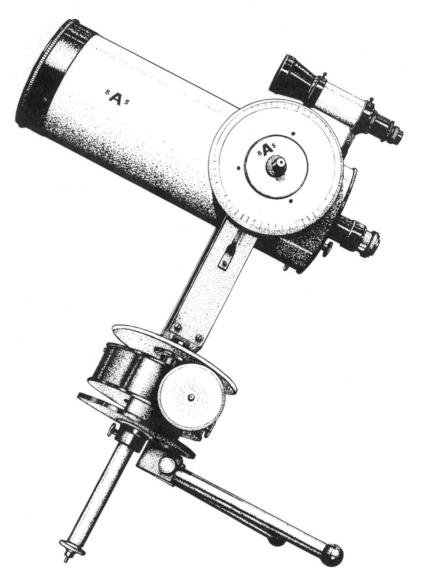

FIG. 15. A Maksutov telescope of 6-inches aperture made by
Bedford Astronomical Supplies of Luton, Bedfordshire. It is
mounted on an equatorial fork and short tripod with slow
motions, circles and an electrical drive. A very compact, portable
instrument of outstanding optical quality.

said to have few rivals: a six-inch telescope which folds into a tube only two feet in length obviously has a strong appeal, particularly when it can produce the same powers of magnification as a refractor of the same aperture but several times its length.

As with the Newtonian and the refractor it is possible to buy the optical parts for a Maksutov telescope and mount them, though perhaps less easily than the mounting of Newtonian optics, in spite of the latter's much greater tube-length.

The Telescope as a Camera

A camera is basically a light-tight box with a hole to admit light to the plate or film which is placed at the rear. In practice there is usually a lens in the hole, though the camera of the old 'pin-hole' variety worked successfully without any lens. The telescope, of whatever form, may look very different from a camera but the

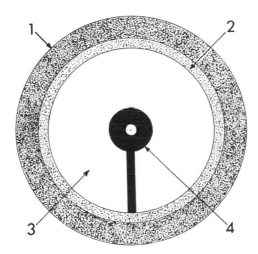

FIG. 16. *View presented by an accurately-collimated Newtonian reflector with eyepiece removed. The flat mirror is supported by a rigid, single arm instead of the more usual three or four-arm spider. 1. Outline of eyepiece draw-tube. 2. Outline of elliptical flat mirror fore-shortened to a circle. 3. Image of main mirror reflected in the flat. 4. Image of flat and its support reflected in the main mirror and then by the flat itself.*

essentials are very much the same, with a lens collecting light and concentrating it into the image which is then examined by the human eye, or allowed to fall on a piece of photographic film or plate. This is photography at the prime focus, but, with ordinary telescopes, the image thus formed is very small. An eyepiece can be used to enlarge it as is done for visual observation.

The amateur astronomer can take photographs of star-trails with a simple box camera which is left with the shutter open for a period of a few hours. True, the result may be merely a few white, curved streaks across the print when it is developed. Many are more ambitious and have taken successful photographs of the moon, planets and star-clusters by the use of their telescopes and a camera of the 35mm. variety. Some highly successful pictures of the sky have been taken with cameras built around old portrait lenses, using long exposures, though in this procedure it must be remembered that the stars are moving; hence the star-trails produced by a stationary camera. Otherwise the camera must be moved to keep in step with the stars.

For photography with a telescope, the reflector is the clear choice, bringing all the light which falls upon it to the same focal point. The lens cannot do this. The visual focus and the photographic (or actinic) focus do not coincide except by the use of the expensive photovisual object-glass.

4 Accessories

The Eyepiece

The object-glass or mirror of a telescope is of prime importance and its most costly element, but its performance is still very dependent upon some other parts. The secondary mirror, or flat, in the case of the reflector, must match the optical quality of the primary mirror. Many a good main mirror has been condemned in haste when the fault has lain with the flat and the result may still be unsatisfactory if they are not matched by the performance of the eyepiece. Some years ago the reflector was often roundly condemned for what was alleged to be inferior definition, though in many instances this was almost certainly due to the application of poor eyepieces, or eyepieces of a type not suited to the short focal length and hence steeply convergent cone of light reflected from the mirror.

The eyepiece can be used in the form of a simple convex lens, or a concave lens as Galileo's telescope had and such as was used more recently in small opera glasses. An eyepiece of this kind has, in fact one important advantage. Less light is lost in passing through one piece of glass than two, three, four or six which may be found in some designs. Otherwise the single lens is not to be recommended. It will have a very small field of good definition. Immediately an object slips from the centre of the field there will be intolerable distortion.

With a reflecting telescope, especially the Newtonian, it is essential to equip the instrument with eyepieces of suitable types. The refractor, largely on account of its greater focal length, is less demanding in the matter of design, though it goes without saying that a bad eyepiece will be bad with any telescope

A large number of eyepieces is an unnecessary extravagance. They have now become expensive items and the time when there were

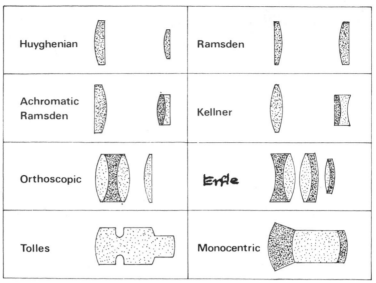

FIG. 17. Types of Eyepieces. With a telescope of long focal
length, such as the refractor, all are able to give a good performance. With the Newtonian of normal f/6 to f/8 focal ratio
the Huyghenian is unsuitable.

large numbers of highly-corrected ones taken from ex-service equipment and available very cheaply, has gone. There is much to be said
for having three or four of varying focal lengths to provide a low
power and wide field, a medium power for use on the moon and
planets and a high power for occasions of exceptional 'seeing' conditions or for tackling close double-stars. The low power should
be one which gives a magnification of about 10 or 15 for each inch
of aperture; the medium power should give about 25 or 30 per inch
and the high power perhaps 50 per inch. It is sometimes asserted
that a first-class lens or mirror ought to bear a magnification of
100 per inch of aperture, but, as a general rule, this is far too high.
Beyond a power of 50 to the inch it will be found that nothing
is gained, since the image of a planet soon becomes ill-defined and
'washed-out'. The fact that Sir William Herschel is said to have used
a power exceeding 6,000 times with one of his telescopes must be
regarded as the exception which proves the rule; the rule being that
experienced observers are content with moderate powers which do

not reduce the brightness of an object seriously, give sharper definition, a wider field of view and make it easier to follow their target
as it moves across the sky.

The main eyepiece designs are as follows:

The Huyghenian is the commonest kind of eyepiece. It is composed of two elements and is sometimes referred to as a negative
form. The image is formed between the two components and it
cannot be used as a simple magnifying-glass for use on its own.

It is among the cheaper types and works quite effectively with
mirrors or object-glasses of long focal lengths, where the cone of light
is comparatively narrow. For short-focus telescopes of any kind it is
quite inadequate except when the appliance called a Barlow lens
is used in front of it. This increases the effective focal length of the
objective and brings the cone of light to a form which can be dealt
with by the eyepiece.

The Ramsden is another relatively simple form, consisting of two
plano-convex lenses with their curved faces towards each other.
For reflecting telescopes of f/6 or f/8 focal ratios the Ramsden is a
much more effective eyepiece than the Huyghenian. It is sometimes called a positive eyepiece. The image plane lies in front of
the field lens(farthest from the eye) and if cross wires are needed,
as in the case of a finder telescope, the Ramsden is the form required.

The Kellner bears some resemblance to the Ramsden in design.
The eye-lens (nearest to the eye) is an achromatic combination of
two elements so that it is more highly corrected than the Ramsden
and will perform well with all telescopes. It is widely used as an
eyepiece for binoculars and many of those used in service equipment were of this type.

The Orthoscopic is a still more complex and highly-corrected form
of eyepiece and therefore more costly to buy. It was first introduced
by the famous German optical firm of Carl Zeiss. It gives excellent
results with any type of telescope. The field of view is wide and the
eye-relief is greater than most others allow. This is of obvious importance to spectacle-wearers and a convenience to everyone else who
uses a telescope. There are a number of variations but all have
several components.

The Tolles has been referred to, a little slightingly perhaps, as 'the poor man's orthoscopic' though this rather patronising epithet scarcely does justice to its merits. It is made from a small cylinder of glass which is convex-shaped at both ends. It is certainly less costly to make than the orthoscopic. It provides excellent definition and freedom from 'ghost' images, those irritating wisps of light which, with some eyepieces, can mislead the beginner into the belief that he has stumbled upon a new comet, perhaps an incipient invasion of the earth from outer space or an extra satellite for Jupiter.

The Erfle has earned a glowing reputation in some quarters as the 'Rolls Royce' of eyepieces, though it was not originally diesigned for looking at the sky except in its war-time role as part of the army's anti-aircraft equipment. Its six components give a wide field of view with exceptionally fine definition and colour correction, though obviously some light is lost with more air-glass surfaces to scatter the rays by reflection and absorption in the lenses. There is good eye-relief with the Erfle too, but its cost is likely to be prohibitive. Some years ago the eyepiece was available in the small elbow telescope used by the services for no more than a few shillings, but the ones which appear on the market now cost something in the region of £20 or $40. A few are being made specially for astronomical work but they must be considered as luxuries.

The Monocentric is undoubtedly the best eyepiece yet designed for moderately high magnifications. It is a cemented triplet combination. Though with a smaller field of view than the Erfle and the Orthoscopic, it has good eye-relief, freedom from 'ghosts' and other aberrations. Very few are available and those are very expensive.

The Barlow Lens

This is a useful accessory whatever eyepiece the observer has and it is particularly valuable with telescopes of fairly short focal length. It is a negative achromatic lens of two components, mounted in a short piece of tube which slides through the telescope's draw-tube in front of the eyepiece. In this way it intercepts the column of light before it gets to the eyepiece, replacing the original cone of rays by one which is more narrowly convergent, as shown in the

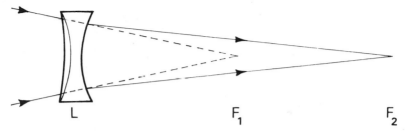

FIG. 18. The Barlow lens intercepts the cone of light-rays from the object-glass or mirror and narrows the angle of the convergent rays, extending the effective focal length and, in consequence, the magnification of the eyepiece. F1 marks the original focus and F2 the modified focus produced by the action of the Barlow lens.

diagram. This has the effect of throwing the focal point further out and increasing the magnification of the eyepiece by a factor which varies with the exact position of the Barlow. The power of the eye-piece may be doubled or trebled at the cost of a small loss of light. An additional merit of the Barlow device is that it enables the observer to make use of eyepieces of relatively long focal length and good eye relief while achieving sufficient magnifying power. Some forms of eyepiece are very awkward to use in short focal lengths, the eye-lens being very small and requiring the eye to be placed very close to it. With the Barlow lens it is possible to make use of the Huyghenian form of eyepiece on a Newtonian reflector or other short-focus telescope with which the Huyghenian alone would be very poor in performance.

The variable power obtainable by the use of the Barlow lens has been exploited in the type of lens described as a 'Zoom' eye-piece. With this the position of the Barlow can be altered by an adjustment which is placed on the outside of the draw-tube. Otherwise the eyepiece draw-tube has to be removed from the focussing mount in order to slide the Barlow in or out as required. Separating it from the eyepiece increases the Barlow's amplification.

Other Accessories

These include such items as the star-diagonal eyepiece which has already been mentioned in connection with refractors. It serves no useful purpose with a Newtonian since it is simply a device for overcoming the awkward observing position of the refractor and, of course, the Cassegrain, which also has the eyepiece at the lower end of the tube. The sun-diagonal, or solar wedge, can be a convenience for observations of sunspots. Its surface reflects only a small fraction of the sun's light to the eyepiece, the rest of the radiation, including the dangerous heat, passing harmlessly out of the telescope without coming near the observer's eye. For terrestrial viewing the Newtonian is not the most desirable form of telescope, but the refractor and the Maksutov are often fitted with erecting devices in the form of the terrestrial eyepiece which usually contains two extra simple lenses to invert the image, or a more compact prismatic erector.

The finder telescope, which is normally fitted to even quite small commercially-made refracting telescopes and reflectors, is an essential if the observer wishes to avoid the trouble and loss of valuable observing time involved in directing an instrument to the target. It is simply a smaller version of the refracting telescope, with a low magnification and wide field. Cross wires, or some other form of graticule, are usually inserted in the eyepiece but, whether they are or not, it is important that the finder should be correctly adjusted by its retaining screws so that an object in the centre of the field is seen in the smaller field of a high-power eyepiece in the main telescope.

The Focussing Mount

The small terrestrial telescope is normally focussed by a simple sliding tube arrangement, but this is not really adequate for astronomical work where a very fine adjustment is needed. A smoothly-acting system is essential. This can be supplied by the rack and pinion with an inner sliding draw-tube. Some of the best examples of this type have two controls, one of which gives a coarse adjustment and the other a fine one, somewhat similar to those fitted on good quality microscopes. A spiral focussing tube is capable of very fine adjustment and, in some models, has been combined with the rack and pinion. But, whichever type of focussing mount is used, it is desirable that the diameter of the draw-tube should be wide enough to accommodate a fairly large, wide-field eyepiece.

Mounting the Telescope

A telescope even as large as a 12-inch Newtonian *can* be used by simply resting the lower end on the ground and the body against a large box or the back of a garden seat. In this way, with a square wooden tube and an ex-army elbow telescope as a finder, I have had some excellent views of the moon, planets and stars. It is, however, not a procedure to be recommended, except perhaps to the amateur telescope-builder who has yet to acquire the proper form of mounting. For this, the essentials are that the mount, of whatever design, should hold the tube securely while allowing it to be moved freely to any part of the sky. This sounds comparatively straightforward, but in fact a vast amount of thought and ingenuity has been devoted to working out the best form of construction to do it. As usual, there is no perfect design, each having its quota of good and bad points. Personal preference is also involved.

The Altazimuth

As its name implies, this form of mounting permits movement of the telescope in the two planes of altitude (vertical) and azimuth (horizontal). It can, therefore, be directed to any part of the sky and terrestrial telescopes invariably have this form of mount. If the telescope is a small refractor the mount may be a metal pillar in which the altitude movement is supplied by a single joint below the tube near its point of balance. This often accompanies a small table tripod, the pillar and claw mounting, which has very little to commend it. The main objection is its unsteadiness, though this is sometimes improved by the addition of one or more steadying-rods from the eyepiece end of the tube to the base of the tripod. The writer's Dollond 3-inch refractor is fitted with a pair of tele-scopic steadying rods of this kind, but although some vibrations are avoided by it, the movement of the telescope in a horizontal plane is very limited without constant adjustments of the rods.

A much more satisfactory form of altazimuth is that in which the telescope is supported on opposite sides of the tube by short, stub axles which swing in trunnion arms on top of a tall, rigid garden tripod. Here, too, the fitting of steadying rods is an additional benefit. In a long and highly informative chapter in a well-known volume of half a century ago, Hutchinson's *Splendour of the Heavens,* the late M.A. Ainslie, writing on 'The Amateur at Work' recalls his experience with a 4-inch refractor fitted in this way without any other refinements and found it a good type of mount. My own Cooke 5-inch photovisual refractor was mounted in this way, but with slow motion controls as well and this proved very rigid in support and smooth in action.

As applied to the Newtonian reflector the altazimuth follows the same principle, but the tripod or pillar which carries the mount is shorter and a steadying rod attached to the top end of the tube normally incorporates a slow motion movement in the vertical plane, operated by a disc or wheel which is conveniently near the observer's position at the eyepiece.

The Equatorial

The major disadvantage of the altazimuth mounting lies in the necessity to move the telescope in both planes simultaneously to follow a star or planet which, through the Earth's rotation, is tracing a curved path across the sky. All forms of the equatorial mount are designed to eliminate this inconvenience. The altazimuth mount can, in fact, be converted to equatorial motion by tilting it so that the main (vertical) axis of the mount points to the celestial north pole, which, for all but the most exacting requirements, is indicated with sufficient accuracy by the Pole Star. The axis, in addition to being elevated to the correct altitude, i.e. that of the celestial pole, must also be set fairly accurately in the meridian. The greater the accuracy of its initial setting the more efficient it will be in use. This really means that a permanent observing site is desirable. An equatorial mounting which has to be erected and dismantled at each observing session, will lose a good deal of its advantage over the altazimuth.

The equatorial is also a necessity for the observer who needs a mechanical or electrical drive for the telescope, whether for visual work or photography, or wishes to use graduated circles for locating objects invisible to the naked eye.

One of the commonest forms of equatorial mounting is that first invented by Joseph von Fraunhofer (1787-1826), the German optician and astronomer who gave his name to the dark lines in the spectrum bands of the stars. The telescope is mounted at one end of the declination axis, which allows the tube to move in altitude (or declination). The main polar axis joins this near the mid-point, and a weight, or series of weights, attached to the other end of the declination axis acts as a counterbalance to the telescope. The mounting is fairly compact in size but, in certain positions, the pillar of the mount restricts the movement of the telescope which must then be swung round through a half-circle and brought to the other side of the pillar. If the telescope is a small refractor the German mounting can be portable, but, if its aperture is 4 inches or larger, the task of attaching it to the flange or cradle at the end of the declination axis and its removal later in the dark, can be a difficult and trying operation. For a portable instrument the trunnion mounted altazimuth is, unquestionably, a more convenient design.

The Fork

If the trunnion arms of such an altazimuth are extended in length and the legs of the tripod are shortened, by tilting the main axis towards the celestial north pole we have a more useful form of mounting for a Newtonian reflector, the fork. The style is not suited to the refractor because of its long tube, but the tube of the reflector should be able to swing right through the fork otherwise the part of the northern sky around the pole becomes impossible to reach. On the whole this may not be of great importance since the sun, moon and planets are never in that direction. The altazimuth, the German equatorial and the fork type by no means exhaust the list of possibilities, but they are the three basic types most commonly found. Some others have been devised which are variations of the basic designs but while they gain some minor advantage in one respect they introduce problems of their own. The Springfield mounting is one of these. It has the advantage of a fixed eyepiece position but it involves a mirror-image reversal which many observers dislike and there is also a third reflection with the unavoidable extra loss of light.

Among amateur telescope constructors there have been many ingenious examples of mounts made from the most unpromising materials and odd components discovered in scrapyards and among broken machinery. For the amateur who is a skilled metal-worker and who has access to a lathe, the possibilities are extensive not to say economical, for professionally-made mountings, like all telescope equipment, can be inordinately expensive.

Drives and Circles

These are applicable only to the equatorial form of mount and whether they justify the additional cost or the trouble of making them will depend upon the ambitions of the individual observer. A slow motion which is driven manually is quite adequate for ordinary observations. This is usually provided by a large worm-wheel attached to the polar axis and turned by a rod which has a universal joint where it is attached to the worm so that it can be operated at varying angles to suit the observer's position at the eyepiece. Some form of clock was often used in the past to accomplish the driving so that the observer could devote all his attention to the object in view. The clock was usually weight-driven but some spring-operated motors have been used and even the motors from old, hand-wound gramophones. Today the electric drive has replaced the mechanical clock but it should be emphasised that an efficient drive is wasted on anything less than a well-mounted instrument housed at a permanent site and accurately adjusted as to the observer's latitude and meridian.

Setting circles are considered to be a convenience rather than a necessity and this probably sums the matter up fairly. If the observer wishes to find a planet in the daylight sky it can normally be done only by the use of circles, plus, of course, a knowledge of the object's position in the astronomer's two co-ordinates of right ascension and declination. The latter is the angular distance of the object measured north or south from the celestial equator (as latitude is measured from the terrestrial equator). The right ascension coordinate is measured in hours and minutes and is the angle between the object and the First Point of Aries where the ecliptic crosses the celestial equator. The measurement of R.A. takes place in an eastward direction round the celestial equator. In the case of the planets the position is also identifiable by reference to the ecliptic near which their paths lie and the measurements in this case are called latitude (instead of declination) and longitude, measured in degrees from the First Point of Aries eastwards as for R.A. but along the line of the ecliptic instead of the celestial equator.

Apart from day-time searches, the use of circles, so far as the amateur astronomer is concerned, is probably little more than an interesting technical exercise and no observer need feel unduly deprived without them.

Provide the tube of the refractor with a dew-cap and the finder telescope as well. If you use a portable reflector cover the main mirror before bringing it indoors. Otherwise dew will form on it very quickly. If this happens let it dry off but do not wipe it with a cloth.

5 Testing the Telescope

For ordinary terrestrial use the assessment of a telescope's quality is a fairly straightforward matter. If the focussing action is smooth and the image comes briskly to a sharp focus showing good definition there can be little wrong with it. In the case of telescopes for astronomical use the criteria are more exacting since the demands on the instrument are greater. This applies to astronomical telescopes of all kinds. We need not ask for, or expect, perfection, but the requirements of the optical components are now so well understood and procedures of manufacture and testing have become so highly perfected that the production of a first-rate objective, whether a lens or a mirror, is now a fairly routine process. Nevertheless, there are optical parts to be met with which fall short of the accuracy needed, just as there are on the market binoculars, new and second-hand, which show faulty collimation. The observer will be well advised to make himself familiar with the few simple tests which enable us to judge the performance of a telescope with some assurance, and, if a fault is detected, to understand how it may best be put right.

The Rev. T.W.Webb has some pertinent things to say on the subject in the first chapter of *Celestial Objects*. Appearances — that is to say externals — may be misleading. 'Inferior articles may be showily got up, and the outside must go for nothing'. He goes on to remark that the quality of a lens, or mirror, cannot be assessed by its surface polish nor the clarity of the glass. Even though some scratches or air-bubbles may be present they do not necessarily rule out the possibility of a good performance. 'Actual performance is the only test' he concludes.

This is undoubtedly true, but an accurate judgement demands a modicum of experience. The process is a good deal more subtle than simple measurement and even experienced people are prone to personal vagaries and prejudices in the matter, such as, for example, tolerance of the false colour in the image produced by a refractor's object-glass. In any case, if the mirror or object-glass has already been used, one should resist the temptation to take out a handkerchief or duster and attack the accumulation of dust or dirt with a hearty rub. Such treatment will only damage what may be an excellent objective, or make a mediocre one bad. Neither can an accurate opinion be formed by taking the telescope out to the garden, putting in the highest powered eyepiece — usually the shortest one — and looking at the moon or, even more unwisely, at the planet Venus if it is visible.

Any telescope taken outside into the cold air from normal room temperature must be allowed to cool down. This applies more especially to the open-tubed Newtonian. An observer who fails to follow this simple precaution, and jumps to an immediate conclusion from the appearance of the image, is doomed to harsh disappointment. Even after the telescope has had a chance to cool to the temperature around it there may still be atmospheric turbulence and judgement as to how far this is responsible for an indifferent image is something which requires an experienced eye. Choose, if possible, an evening of fine, settled weather with little wind. If there is much breeze it will severely test the telescope's mounting and some tremors are unavoidable.

Do not expect a clear, sharply-defined image of a planet when it is low in the sky. Its light is dimmed and subject to the turbulence of the lower atmosphere.

The test object should be examined when near the meridian, or at a fair altitude above the horizon east or west of it. Venus, as Webb says, is a severe test against a dark sky, on account of its brilliance and no experienced observer would expect a fair verdict from a glance at that brightest of stars, Sirius, low in the southern winter sky. Against a light background of sky, Venus presents a more tolerable target though it rarely provides an absolutely steady image, and, if any markings are suspected on its bland, white surface, they are probably only in the eye of the observer, like the brilliant colours which a friend of Denning's became used to seeing round Jupiter in his small refractor. When he found they were missing when viewed in Denning's reflector he seems to have resented the suggestion that the coloured fringe existed only in his own telescope.

Jupiter and Saturn will provide good tests of a telescope's defining powers, especially the details of Jupiter's cloud-belts and the division in Saturn's ring, though the latter may be difficult for the beginner unless the rings are near their fully open phase.

A star of moderate brightness will also provide a reliable test under reasonably good weather conditions. It should come into and out of focus sharply, with the small 'spurious' disc clearly defined and surrounded by no more than one or two faint diffraction rings as the diagram shows. In a reflector, however, there will be some rays of light surrounding it which are caused by the arm which supports the flat mirror or prism. A single arm will produce a double ray. A three-armed 'spider' supporting the flat will show six rays. These effects of the flat supports are most noticeable with bright stars.

Making Your Own Star

In a climate where cloudy skies predominate it can be a severe test of patience waiting for a suitable opportunity to try out a telescope on the sky. The clouds will eventually roll away, but a night of exceptional clarity does not always produce ideal 'seeing'. As an alternative to the real stars we can make an artificial one with the help of the sun. The test can then be carried out in greater comfort during daylight hours. An object with a sharply curved, reflective surface will project an image of the sun which is a tolerable substitute for a real star. A spherical glass flask, a silver ball which normally decorates a Christmas tree, or a thermometer bulb with the sun's image reflected in it, will serve the purpose admirably. On a cloudy night, or when the atmosphere is otherwise unfavourable, an illuminated pin-hole in a sheet of cardboard or metal will do, the light being supplied by a lamp or torch behind the pierced screen.

The home-made 'star' should be set up at some distance from the telescope, necessarily beyond the point at which the telescope's objective is focussed when the draw-tube is fully extended.

A low power eyepiece should be used and the 'star' brought to

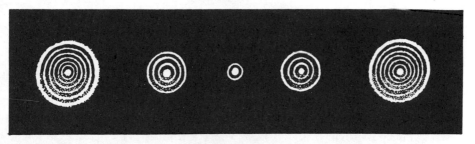

REFRACTOR STAR IMAGES

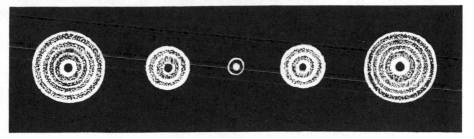

REFLECTOR STAR IMAGES

FIG. 19. It is possible to make an accurate assessment of the optical quality of a telescope from the appearance of the out-of-focus images presented by a star. The star can be produced artificially, by using the image of the sun reflected in a thermometer bulb, for example. A high quality telescope should produce a sequence of images similar to these when the eyepiece is moved from a point outside focus (left) to a similar distance inside focus (right). The dark central spot in the reflector image is caused by the flat mirror.

the centre of the field of view. If the eyepiece is now removed, and the draw-tube pushed in a short distance, it will be possible to discover some things about the object-glass which are not normally visible. If the illuminated field is uniformly bright all over, there are no flaws or striae, streaks of uneven density in the glass. If there are some present they will be seen, as will also any particles of dust, patches of dirt and, possibly, finger-marks.

If the artificial star is now examined with an eyepiece of higher power, when carefully focussed it should exhibit the same form as a real star does on a night of good 'seeing'. This is the small disc and rings in which the latter should be perfectly round and concentric with the disc. If there is any deformity or lack of symmetry about the image, something is wrong with the telescope. This may be seen even more clearly if the eyepiece is either pushed in, or drawn out, beyond the focus, when the small disc and rings will expand into a series of larger, diffraction rings. These, too, should be nicely concentric and of fairly uniform brightness. If there is any departure from the circular shape or any conspicuous irregularity of form or brightness, there is a fault present.

In fact, the out-of-focus star-disc, whether of the natural or artificial variety, and whether viewed with a refractor or a reflector, represents an optical test of considerable delicacy, especially as regards the similarity of appearances when the eye-piece is moved in and out for equal distances on each side of the focal point. Ideally they should be exactly alike on each side of the focus. An accurate assessment of the objective's quality from this test is something that requires a certain amount of experience, and knowledge of what to look for. Such faults as may appear are not necessarily indicative of a poor lens or mirror. The trouble may be no more than an error of collimation which can be corrected by a touch of the adjusting screws. It may lie with the eyepiece, so that more than one should be tried. In the Newtonian reflector it can, and often does, lie with the small flat secondary mirror. Producing a really flat surface calls for optical skill no less than that required to produce an accurate paraboloid.

6 Observing the Sun

'The solar phenomena' remarks Webb, 'are especially wonderful' and he goes on to speak of 'the unrivalled pre-eminence of that glorious sphere', while warning of the dangers involved in its observation. The sun's rays are of dazzling brilliance to the unaided eye, except when dimmed by the lower atmosphere near the horizon. When the rays are concentrated by the objective of even a very small telescope, their effect upon the human eye could be devastating. It is said that Galileo, and even the experienced William Herschel, inflicted serious damage upon their sight by inadequate precautions when looking at the sun.

The refracting telescopes of three or four inches aperture, which were produced some years ago, almost invariably had a set of Huyghenian eyepieces each fitted with a 'sun-cap' of dark glass intended for use in observing the sun. It was seldom indicated that the use of these eyepieces for direct viewing of the sun could be undertaken without risk only when it was low in the sky, or already substantially dimmed by mist or thin cloud.

Many activities involve an element of risk and are still undertaken in the full knowledge of the possible consequences, but it is unwise to take such risks, particularly if the penalty may be anything so damaging as partial or total blindness, when a satisfactory alternative method of procedure is available with complete safety. Such is the method of solar observation by the means of projecting the sun's image on a white screen. The only safe alternative is to use the sun-diagonal eyepiece already mentioned in the section dealing with accessories. Even in this case, however, the use of a dark sun-cap on the eyepiece is essential.

If a telescope is being used on the sun it is wisest to 'stop down' the aperture to about 4 inches, since the problem here is the abundance of light rather than the lack of it. This procedure is very familiar to the photographer, though the astronomer has no conveniently-operated iris diaphragm. A simple cover of cardboard can be arranged with a central aperture of the appropriate size, say three or four inches in diameter. This can be attached to the eyepiece end of the telescope, mounted on a light support, and the sun's image can be brought to a focus on the screen by the sliding draw-tube or rack and pinion. It is advisable to protect the screen from the direct rays of the sun and, better still, to arrange a partially enclosed box around the screen with a side aperture to permit observation of the sun's image. Even with a small telescope there will usually be sunspots visible and, with a higher magnification, it may be possible to see the granular 'rice-grain' structure of the solar surface.

Sunspots appear in a variety of shapes and sizes. Even the small ones, which are only just visible from the Earth, are several hundreds of miles in diameter, while large groups are sometimes seen which cover an area of the sun's surface amounting to many millions of square miles. Such was the size of the immense group which appeared in April 1947. Occasionally a single spot is big enough to be seen with the unaided eye when the sun is rising or setting and when higher in the sky if the light is screened by a very dark piece of glass. The sunspot cycle of a little over eleven years (an average time) was first detected by the remarkable patience and persistence of Heinrich Schwabe of Dessau, Germany in 1843, after some forty years of routine observations.

Some solar observers make use of specially prepared discs which are marked with the sun's lines of latitude and longitude. These are transparent and by fitting them over the projected image the position of the sunspots can be accurately recorded. The sun, it should be remembered, is rotating on its axis about every 24 days though this figure varies, unlike the rotational period of the Earth, since the sun is a sphere of gas. The solar regions beyond the equator take longer to rotate, reaching a maximum of 34 days at the poles. Also the solar axis of rotation is not always constant with reference to the Earth, so that the observer who uses the 'Stonyhurst Discs' for sunspot recording has a series from which to choose the appropriately-marked disc of the day.

At the commencement of a new upward curve of spot activity

the first indications occur in the higher latitudes on both sides of the solar equator. As the cycle progresses they become more numerous and move towards the equatorial regions and this tendency continues as the downward curve of the cycle takes place. Before they have quite died out, however, the next cycle may be showing itself with the appearance of new spots in the higher latitudes. It is interesting to note that, as often happens in astronomy, like our inability to see the sky 'in depth', the appearance of a sunspot is deceptive. The outer region, or penumbra, has a brownish tinge, but the central core, the umbra, appears black. The blackness, however, is only a matter of contrast. Were we able to isolate the spot from the rest of the sun and put a very powerful searchlight at its centre, the latter would be seen as a black spot against a brilliant background.

That prolific author of astronomical books a century ago, the late R.A. Proctor, in his attractively-written little manual entitled *Half-hours With the Telescope*, describes his 'enhanced feelings of awe and wonder' at the sight of the greatly-enlarged image of a sunspot projected by a star-diagonal eyepiece on the ceiling of a darkened room. 'There', he adds, 'the vast abysm lies pictured; vague imaginings of the vast and incomprehensible agencies at work in the great centre of our system crowd unbidden into my mind; and I seem to feel — not merely think about — the stupendous grandeur of that life-emitting orb'. The evocative prose of such writers as Webb and Proctor, very characteristic of their age, is not to everyone's taste nowadays, though it is a style which not all readers will put to their discredit. As the Irish author, Donn Byrne, once said, there will always be some simple-minded people 'who will stupidly go on believing that kindness is not begotten by logic, nor heroism a product of carbohydrates'; and astronomy is not confined to cold statistics.

Aside from the changing appearances of the sunspots there are the even more spectacular prominences, huge clouds of glowing hydrogen, arching above the sun's surface at heights of many thousands of miles. In many cases of the larger prominences there is an association with sunspots. To view them, however, a special form of spectroscope is needed for attachment to the telescope, for which purpose a refractor of about 4 inches aperture is usually considered the most effective, mounted preferably on an equatorial having a mechanical or electrical drive. The operation of the prominence spectroscope is a somewhat delicate procedure and the narrow slit through which the prominences are observed has to be kept in the position of a tangent to the sun's edge. The sun's outer atmosphere, the corona, visible as a glowing, pearly halo, can be seen only on the rare occasion of a total solar eclipse, except by use of a special apparatus, the coronograph, invented in 1930. This works best at a high altitude.

7 Observing the Moon

Although men have walked on the moon, circumnavigated it and taken innumerable close-up and detailed photographs, the appeal of our satellite for the beginner in astronomy, and even for the observer of long experience, remains unimpaired. To quote Webb again, 'Many a pleasant hour awaits the student in these wonderful regions', although the innocent fancy of the islanders of Teneriffe a century ago that the telescope of a visiting astronomer might show them their favourite goats on the surface has doubtless, long since evaporated.

The smallest telescope, or even a pair of binoculars will show some of the larger lunar features among the craters, mountain ranges and vast, walled plains. It is, however, essential for the observer to seek them at the appropriate time. The moon's atmosphere is so thin as to be negligible and the sunlight is unsoftened by any cloud. Along the line of the 'terminator' as astronomers call the curved line of sunrise, between new moon and full, or sunset, between full moon and new, the uneven, rugged landscape is projected in harsh relief by the fierce rays of sunlight.

It is along this margin that the telescope reveals the moon's features in an immense amount of detail, showing up in startling clarity small objects which disappear completely as the shadows shorten with the sun's increasing altitude until, at full moon, the needle-sharp relief of the crescent moon becomes flat and relatively featureless.

In the past history of selenography, as the study of the moon's surface is called, the amateur observer has undoubtedly played an active and honourable role. There have been numerous maps of the surface laboriously compiled and replete with minute detail, such as the one drawn by H.P. Wilkins on a scale of 300 inches to the moon's diameter, as well as many on a smaller scale. More recently the opportunities have been restricted. Some useful work was accomplished in the 'limb' regions, where the moon's libration enables us to see a little way round the edge at various times. This became superfluous from the time when we were able to send spacecraft round the far side and the unknown hemisphere proved to be very much like the one which is turned permanently towards us.

Today, the opportunities for the serious amateur are even more restricted and probably limited to the search for what have become known as 'transient lunar phenomena' or T.L.P. There is obviously very little significant change occurring on the lunar surface. The absence of any appreciable atmosphere means that there is no action by wind and weather as we understand the term. It is, however, highly probable that some small-scale activity is taking place and this has been confirmed by the recording of lunar tremors, like minute earth-quakes, in the seismometer left on the moon by the astronauts of the Apollo XII mission. Emissions of gas have, from time to time, been suspected from certain lunar craters, such as the one which was detected in the Alphonsus formation by the Russian astronomer, Kozyrev in 1958. On other occasions there have been some fairly well-authenticated instances of colour changes. A long-standing controversy surrounds the suggested change alleged to have taken place in the formation called Linné. Before 1840 it seems that most observers recorded and drew it as a crater, whereas for a century past it has been only a very small pit, or craterlet inside a rather diffused, whitish-looking area. In the T.L.P. field there is still an opportunity for the amateur who wishes to be more than a lunar sight-seer.

Many maps of the moon, on a reduced scale, can be found in books like star-atlases, while some very detailed photographic charts are also obtainable. A useful map for identifying the main lunar formations visible in a small or moderate-sized telescope is that prepared some years ago by Elger, who was a former Director of the Lunar section of the British Astronomical Association. This was revised and brought up to date in certain details by Wilkins, producer of the 300-inch map. An attractive and highly informative description of the lunar world will be found in the book entitled *The Moon*, by an American author, G.P. Serviss. It is illustrated by a series of clear photographs, taken on succeeding dates to show the changing phases. There are also some other photographs on a larger scale showing some of the more spectacular lunar landscapes. Unfortunately this volume has long been out of print but may be found in some libraries and, perhaps, in secondhand bookshops.

A reflecting telescope of 6 inches aperture will show a great deal of fine detail on the moon. The smooth, dark areas of the 'Maria' or lunar seas, can be identified easily enough with the unaided eye. A small telescope or pair of binoculars will show craters down to a diameter of about 10 miles, but the 6-inch reveals many more, down to a quarter of this diameter. To get below the one mile barrier a 10-inch is needed but, in every case, there is obviously some degree of uncertainty since the eye is part of the optical system and its powers vary from one person to another.

Occultations of the stars, and more rarely of a planet, can be seen with a small telescope. When the moon passes in front of the star, its light is cut off with dramatic suddenness. These events interest the professional astronomer too, especially in their timing, for there still remain some elements of uncertainty in the moon's movements.

Occultations have been observed ever since the telescope was invented. One of the earliest recorded ones was that by French astronomer, Bullialdus, in 1623. He is remembered also by one of the most spectacular of all the moon's features, a large walled plain of about 40 miles diameter with a complex mountain group in the centre. But if occultation observations are to be of scientific value some highly accurate measurements are essential. It is desirable, for instance, that the observer should know his own location very precisely. This means to an accuracy of about one second of arc, which is a little more than thirty yards on the ground. The observer's height above sea-level is required, to an accuracy of about thirty feet. Equally important is very exact timing. If the amateur decides that he is happy to watch the event solely as a spectacle, then that is understandable. Professional astronomers are now interested in following the moon's progress as it occults some sources of radio waves in its path, but this is a field which lies beyond the instrumental means, and probably outside the active interest of most amateurs.

Vastly more impressive as a spectacle are the occasions when the moon passes into the Earth's shadow cone and undergoes a partial or total eclipse. When this happens, and the sun's light and heat are cut off from the moon, the surface cools very quickly since there is no atmosphere to retain the heat like a blanket. Observations of the lunar surface before and after the eclipse may be of value because it has been suggested that the rapid cooling and then heating of the surface produce visible effects in some formations, including the controversial crater, Linné, in the moon's wide plain called the Sea of Serenity.

8 Observing the Planets

Discoveries of new planets have occurred only at very long intervals. Among the major bodies of the solar system only three have been found, Uranus, Neptune and Pluto, which were not already known to the ancient world, although the small bodies known as the asteroids have been detected in their hundreds revolving in the wide belt between the orbits of Mars and Jupiter. Suspicions have occasionally been aroused as to the existence of other major worlds, inside the orbit of Mercury, the innermost planet, and beyond the orbit of Pluto, the most distant. The planet sought very near to the sun was even given a name, Vulcan, and a close watch was kept on the sun's surface in case the undiscovered planet should reveal itself as a dark spot against the bright globe. The main body of scientific opinion has, for some years past, regarded the existence of Vulcan as a myth, though we can never be completely sure that such a discovery is totally beyond the bounds of possibility.

As with the moon, the serious amateur has found his chances of productive work drastically circumscribed by the advent of the space-probe. No longer do the corridors of science echo to the sound of voices debating the existence of intelligent beings, or even of their canals, on the red planet Mars. If the amateur is content, however, with a more modest role than that of a celestial Columbus, he will find that at least four of the planets — Venus, Mars, Jupiter and Saturn — are rewarding targets for his telescope. Certainly there is keen satisfaction in the mere sighting of the others, the crescent of Mercury in the afterglow of the sunset, 'shining' as Serviss once said 'like a globule of molten metal against the fading curtain'. Uranus, too, presents a very small disc, easily distinguishable from a star, but in neither instance will prolonged gazing with a telescope of the size normally found in amateur hands reveal anything more. Mercury is always low down when in a dark sky and its image seldom steady. Uranus is so remote that a fairly high magnification is needed, and therefore good 'seeing', to detect a trace of the dimly-marked belts which cross the tiny, sea-green disc.

Venus can severely test the optical quality of the telescope after dark, but it can be seen to advantage in a light sky when well removed from the sun. Both Venus and Mercury, lying inside the Earth's orbit, may be seen in transit across the sun. Venus, however, seldom obliges us in this respect. The last occasion was in 1882 and the next will be in the year 2004 followed by another in 2012. After that, a century will pass without one. Mercury's transits occur much more frequently, averaging thirteen occasions per century. They always take place in the months of May or November and the next is due on 12th November 1986.

The exceptional brilliance of Venus is due to the reflective, cloud-laden atmosphere which surrounds the planet. The full moon may give an impression of brightness on a clear night and when high in the sky, but, in fact, its albedo, or reflecting power, is rated at only 7 per cent. In the case of Venus the comparable figure is almost 60 per cent.

The crescent phases of Venus are well seen with a small telescope and some observers have claimed to see traces of vague, dusky markings, but their existence is problematical. The points of the crescent often show a distinctly enhanced brightness though the explanation for this is as yet uncertain.

Mars has, in recent times, yielded up many of its former mysteries to the space-technologists, and the challenge which it presented to the American astronomer Percival Lowell, nearly a century ago, has now lost some of its edge. Lowell built the observatory which bears his name at Flagstaff, Arizona, in 1894, especially to devote his energies to the study of the red planet, and he also began a serious search for a planet which he was convinced lay beyond Neptune. He died in 1916, fourteen years before Pluto was found.

Since Mars tends to present us with a very minute disc, and is, in fact, a relatively small globe, its diameter being just over 4,200 miles, it is not an easy object to observe with a small or medium-sized telescope. A magnifying power of around 300 times, or near that, is needed to show much detail and this requires a steady con-

FIG. 20. The planets at their nearest approach present images of the size indicated in seconds of arc. For the observer using a small or medium-sized telescope it is Venus, Mars, Jupiter and Saturn which take pride of place.

dition of the atmosphere. Very favourable oppositions occur only every fifteen years when the planet is near perihelion, its closest approach to the sun. This is the only time when the owner of a small telescope can expect to see more than a diminutive reddish disc with a few darker streaks. The favourable opposition of 1877 is remembered for the fact that the two satellites of Mars were discovered on that occasion by the American astronomer, Asaph Hall, at the Naval Observatory, Washington. It was also the year in which the famous 'canal' controversy was provoked by the Italian observer, Schiaparelli. Using a refracting telescope of 8 inches diameter he detected various streaks and lines which he described by the Italian word 'canali', meaning channels. Some words in the language can be given a fairly obvious English equivalent without too much distortion of meaning, but Schiaparelli's 'canali' was destined to become a classic example of the hazards inherent in a loose, popular translation when it became 'canals'.

Yet the channels are not wholly mythical, for close-range photographs which have been taken by the Martian probes have clearly shown, besides the craters and rocky terrain, something which looks very like dried-up river beds or other ancient water-courses.

The two satellites are beyond the range of small instruments since they are extremely small bodies. Nothing less than a 12-inch telescope is likely to show these two minute companions to the god of war, so aptly named Phobos and Deimos, Fear and Terror.

The white polar caps may be seen under good conditions with a small telescope, but for really effective views an aperture of at least 6 inches, and preferably more, is needed. With a telescope of this size some useful work is still possible in connection with what may be called the Martian equivalent of T.L.P. on the moon. Mars has an atmosphere, albeit much thinner that that of the earth, and from time to time some form of cloudy obscuration of the surface occurs. Sometimes these have been seen to cover a very large area, as happened in 1956 when the south polar cap was covered.

As with many other celestial targets an experienced eye will see more with a small telescope than the beginner with a large one, but the man with limited instrumental powers may take comfort from the achievements of the German astronomers, Beer and Madler, whose observations of the moon and Mars early last century were highly acclaimed, although they worked with a telescope of less than 4 inches aperture.

There is general agreement among amateur astronomers that, of all the planets, the largest, Jupiter, is the most rewarding for observation with a small or medium-sized telescope. With a very modest magnifying power of 40, its disc expands into a globe equal to that of the full moon seen with the naked eye. The four largest satellites are visible with even less power and their movements, eclipses, transits and shadow-transits, provide an ever-changing panorama of a solar system in miniature, while the globe itself presents variations of configuration and colour in the prominent cloud belts which are matched by no other body in space. Even Saturn, set within its magnificent system of rings, lacks the variety of Jupiter, while the latter, coming to opposition every thirteen months, can be seen in the night sky shining brightly for several months of each year. It is true that summer oppositions are less favourable since the Zodiac belt within which the planets move is then very low in the southern sky. At other seasons, especially when opposition takes place in winter, the giant planet rides high in the southern evening sky among the stars of Taurus, Gemini, or Cancer.

A moderate magnification will show the globe departing markedly from a perfect sphere. So rapid is its rotation, in under ten hours, that the polar diameter is compressed to a value of six thousand miles less than that measured across the planet's equator, while the changing aspect of the disc due to its rotational speed can be readily appreciated in the course of a few minutes. Indeed this very rapidity of change is one of the problems which the serious observer faces, in drawing the belts before their aspect has changed.

As the most distant planet known to the astronomers of the ancient world, Saturn's slow advance among the stars made it the most lethargic and leaden-footed of the celestial orbs; so it became associated with all that was dull and heavy. Unknown to them, the truth was precisely the opposite, for Saturn, though its orbital velocity is relatively slow at six miles per second, has a low enough density to float in water. Its speed of rotation, however, is only a little less than that of Jupiter, taking place in a few minutes over ten hours, and its polar compression is even more marked.

Galileo's small telescope, which magnified Saturn's globe a mere thirty times, gave him only a very imperfect view of the splendid rings and, when their changing phase caused them to disappear from his view, he was completely at a loss for an explanation. In 1610, the year when Galileo made his momentous discoveries in the sky, the rings were already beginning to close and two years later they were invisible. It was not until 1656 that the Dutch observer Huyghens, who gave his name to the well-known eyepiece, saw the globe of Saturn crossed by a dark line. The rings at this time were again closed, but, at a later stage, he saw them clearly and realised that the dark line he had seen a few years before had been the shadow cast by the ring system on the globe of Saturn. The main division in the rings is named after the Italian astronomer, Cassini (1625-1712) who first saw it in the year 1670. Cassini is credited with a number of other important discoveries, including four satellites of Saturn, as well as the first accurate estimates of the rotation periods of Jupiter, Mars and the sun, from observations of their surface markings.

The amateur's three or four-inch refractor is capable of giving attractive views of the planet's main features, including the division named after Cassini, the delicate shading of the belts across the globe — though these are far more static than Jupiter's — and three or four of the satellites. The largest satellite, Titan, is an easy target for even a 2-inch telescope. A 3-inch will capture Iapetus and Rhea, and a 4-inch should reveal two others, Dione and Tethys, all of which were seen by Cassini in the seventeenth century.

Though lacking the spice of variety which makes Jupiter so interesting, Saturn, unquestionably, must be awarded the palm for its sheer beauty, unmatched by anything else in the entire heavens. The late R.A. Proctor, while lamenting that his own four-inch refractor showed little colour on Saturn a century ago, quoted the almost lyrical description given by another observer, Browning, a maker of reflecting telescopes, in which occur such phrases as 'yellow ochre, brown madder, purple madder, orange and purple shaded with sepia, and pale cobalt blue.'

To reproduce the colours on paper, declared the enthusiastic

Browning, we should need to dip our brush in the rainbow, for all colours on earth have a muddy quality compared with those in the skies.

It is likely that had Proctor been using a good reflector, he might have agreed with Browning, for, even allowing for the latter's understandable exuberance at the sight of the ringed planet on what must surely have been a night of exceptional clarity, there is no doubt that the reflecting telescope has a marked superiority over the refractor in the rendering of colour. On Jupiter, Mars and Saturn this advantage is unmistakable when a good mirror is used with an eyepiece of matching quality.

9 Observing the Stars

To the uninitiated, a first glimpse of a star through the telescope can be very disillusioning. Perhaps this was the reaction which Wordsworth saw on the faces of the people he once met in a street where a 'kerb-stone astronomer' had set up his instrument for the entertainment of passers-by. One and all appeared to turn away dissatisfied. It has to be admitted that this reaction is not uncommon, for the viewer's expectations have often been formed by looking at photographs and drawings made with the help of big observatory telescopes, and probably very much enlarged. This frequently happens when a beginner takes his first look at Mars, even though the instrument may be a perfect one and the magnification adequate.

Some people are disappointed to find that the stars look very much the same through a telescope as they do to the naked eye, mere points of light, although there are very many more of them. After a little more experience and understanding of the purpose and potentialities of the telescope in stellar observations, the beginner will find more satisfaction. Using a low-power eyepiece with a wide field of view, he will find pleasure in the impressive star-fields of the Milky Way and the bright clusters such as the famous example in the sword-handle of the hero, Perseus, when it is high in the zenith. For an object in this favourable position, of course, the user of a reflecting telescope scores heavily in comfort and convenience, as he also does in looking at the colours of the stars. Red ones are particularly well seen with the reflector and a notable instance is the star Mu Cephei, in the prominent constellation of Cepheus. This is often called 'the Garnet star' since it was compared to that stone by Sir William Herschel. It is visible to the naked eye, but star colours are rarely conspicuous to the unaided vision except in the cases of the very brightest ones.

Star colours enter very much into the pleasure which many

observers find in looking at the numerous double stars to be found in every constellation of the heavens. Sometimes the colours owe their effect to the contrast between the two components of the pair. The star Beta Cygni, in the well-marked constellation of Cygnus, the Swan, or the Northern Cross, is an outstanding double star for any small telescope. The components are widely separated and the colours are gold and blue. In the same constellation is another easy double star, 61 Cygni, the nearest star in the northern hemisphere of the sky and the first to have its parallax measured, in the year 1838 by the German astronomer, Wilhelm Bessel.

Many catalogues of double stars have been made for they are numbered in thousands. Webb's *Celestial Objects*, in its later two-volume editions, has one entire volume devoted to these and other stellar objects, clusters, nebulae etc. Double stars are often used by amateurs to test the resolving powers of their telescopes, or the light-gathering power where one component is very faint. The Pole Star is a popular target, and a fairly easy one except with a very small telescope and an inexperienced eye. The brilliant Vega, leader of the constellation of Lyra, the Lyre, has a faint companion which is difficult to see with small instruments as is also the companion to the red star Aldebaran, in the constellation of Taurus, the Bull.

In using such objects to test the powers of his telescope, always combined, let it be remembered, with the powers of his own eye, the observer must not overlook that the use of an old catalogue, such as that of Webb, may be misleading. In some instances the figures given for the separation of the stars (in seconds of arc) and the position angles, (in degrees from 0 to 360 measured anti-clockwise from the North, the 6 o'clock position with a normal inverting astronomical eyepiece) will be out of date, for the stars, in many cases, will have changed their relative positions considerably after half a century or more. Up-to-date values are given in some more recent lists, such as the one which appears in the observer's annual of the British Astronomical Association.

If the amateur has serious expectations of working in stellar astronomy he now has very little scope outside the field of long-period variable stars. Some of these can be studied with small telescopes or binoculars and the British Astronomical Association has a section devoted exclusively to co-ordinating the work of amateurs in this direction. The variables of shorter period, such as the Cepheids, are given close attention by the professional astronomers owing to their importance in the determination of the distances of stars and galaxies by the application of the period-luminosity law. A telescope of six or eight inches aperture will bring a vast number of long-period variables within the observer's compass. Apart from these, however, the amateur may consider himself a spectator, though a privileged one, for the visual charm of the night sky is inexhaustible.

Many people have spent years of their lives, and vast fortunes, in travelling about this one small planet, while missing entirely the wonders which they could have seen in one clear evening from their own back-garden, with nothing more than a small telescope, a pair of binoculars or even their own unaided eyes.

Appendix

In the days of the elder Herschel, and for some considerable time afterwards, astronomical telescopes were available only to those with a great deal of money. This was one reason why Herschel himself turned to making his own telescopes, an example which has been successfully followed ever since. In the eighteenth century the celebrated London maker, Tulley, sold his 10-inch Newtonian reflectors for £315 which, at that time, must have represented a very substantial sum. At the end of the last century the Rev. T.W. Webb's 9-inch reflector was sold for as little as £6. Although small, brass-tubed refractors could be bought comparatively cheaply a few years ago, the situation has since changed radically. My Dollond 3-inch, on a pillar and claw stand with a set of eyepieces and a fitted mahogany case, was bought for £25 as recently as 1974 and has an interesting association with the former Poet Laureate, Robert Bridges. Today these telescopes are beginning to appear in the London sale-rooms and in antique shops at much-inflated prices, while collectors are searching for the original brass pillar and claw tripods for telescopes whose former owners heeded the advice of practical astronomers and discarded them.

There are, however, many small Japanese refractors available, though those of good quality are not cheap. It is, after all, unreasonable to expect frequent bargains in optical equipment which demands a high standard of craftsmanship in its production.

Old Dr. Kitchiner deals with the matter of bargain-hunting in his typically amusing way with what he describes as 'The Anecdote of the Humorous Hosier'.

A shopkeeper displayed a notice which announced 'Cheapest and Best Stockings'. To enquiries after the goods he replied. 'Here is a pair of each. These are a shilling; those are a guinea a pair'.

Kitchiner goes on to warn his readers that 'the quality of telescopes varies in quite as many degrees as that of stockings'. Having experimented with more than fifty telescopes over a period of thirty years, at a cost of £2000, the author came to know the subject, and the market, well.

As to how much saving can be achieved by self-help is a matter which must depend largely upon the skill of the individual, although mirror-making has been said to demand care and patience rather than a high level of craftsmanship. In the United States, particularly, amateur telescope-making has reached a remarkably advanced and sophisticated level and instruments of outstanding quality to a wide variety of designs, have been made there.

The amateur who feels that he is unable, or unwilling, to face the task of mirror-making, can still save a good deal of the cost of a complete telescope by buying a decent set of optics and doing the mounting himself. Any of the suppliers listed can be relied upon to provide mirrors of moderate, or even large, sizes with a good 'figure', together with cells, flat-mounts, finder telescopes and eyepiece mounts. The work of putting these into a simple, square wooden tube is not something which demands exceptional skill and the same may be said of making a fairly basic type of mount. Slow motion controls and a mechanical or electrical drive, are matters which obviously require more knowledge and expertise.

In most parts of Britain and America there are astronomical societies with experienced amateurs who are usually very willing to assist the beginner.

Suppliers in Britain

Brunnings (Holborn) Ltd. 133 High Holborn, London
Charles Frank Ltd. 144 Ingram St, Glasgow, Scotland
Astronomical Equipment Ltd. Unit D, Lea Industries Estate, Ox Lane, Harpenden, Herts.
Fullerscopes Ltd. Telescope House, 63 Farringdon Rd. London
H.N. Irving & Son Ltd. 258 Kingston Rd. Teddington, Middlesex
David Hinds Ltd. Optical Works, 32 The Mall, Ealing, London
H.W. English, 469 Rayleigh Road, Hutton, Brentwood, Essex
Bedford Astronomical Supplies, 5B Old Bedford Road, Luton, Beds.
Tel-Optics, 13 Park Avenue, Winterbourne, Bristol
Revor Optical Ltd. 12-13 Henrietta St. London

Suppliers in the United States

Cave Optical Co. Long Beach, California
Celestron Pacific, Gardena, California
Criterion Manufacturing Co. Hartford, Connecticut
Edmund Scientific Co. Barrington, New Jersey
Jaegers A. Lynbrook, New York
Questar Corporation, New Hope, Passadena
Star-Liner Co. Tucson, Arizona
Unitron Scientific Inc. Newton Highlands, Mass.
University Optics, Ann Arbor, Michigan

97

Bibliography

Since the scrolls of Ptolemy's ancient classic, *The Almagest*, were rescued from the ruins of the Alexandria library nearly two thousand years ago, the literature of astronomy has gradually expanded with the universe and at an increasing rate during the last century. If the amateur astronomer merely wishes to have a comprehensive, authoritative and up-to-date account of the science he can obtain it in one volume for the sum of fifteen pounds. This is the recently published *Cambridge Encyclopaedia of Astronomy* edited by Dr. Simon Mitton of the Cambridge Institute of Astronomy. There are, however, some other aspects of the subject beyond the purely descriptive, and there are many books, apart from the acknowledged astronomical classics. Sir John Herschel's *Outlines of Astronomy*, is one among many which have long been out of print, but which turn up now and again in secondhand bookshops. These, like old scientific instruments, are now being collected in their own right and their prices have reacted accordingly.

The browser in the antiquarian and secondhand market will not get the latest results of modern research, but he may be fortunate in finding other things besides a version of astronomy which is, in some respects, dated. There are the elegantly bound school and college prize books, for example, like the three volume Clarendon Press edition of Chambers' major work, *A Handbook of Descriptive and Practical Astronomy*, dated 1891, which I came across in a Cambridge book-seller's, bound in richly-tooled leather and emblazoned on the front with the arms of Exeter College, Oxford. Inside the cover of the first volume was a letter, in a spidery hand, dated April 1903 from Norwich, Conn. America, and signed 'Levi W. Meech, Actuary', giving in some detail the writer's conclusion, calculated by reference to the position of the sun and the ecliptic, that the creation of Adam occurred in B.C. 3958. Not, perhaps, an earth-shaking pronouncement, but I know a former Oxford University 'scout' who looked after the rooms once occupied by Einstein during a stay at Christchurch College, Oxford, and now remembers with some considerable dismay how dutifully he disposed of the contents of the great man's waste-paper basket, and other papers, some of which would, no doubt, have considerable interest (and value) today.

A few years ago I came across a copy of the late H.H. Turner's book entitled *Astronomical Discovery*, which includes a detailed account of the finding of Neptune and the events which preceded it. At the time of his book's publication Turner held the Savilian Chair of Astronomy at Oxford. This book also contained two letters of much interest, with particular reference to the Neptune episode. One was addressed from Flagstaff Observatory, Arizona, in the hand of Professor Percival Lowell, and the other was from a member of the Adams family, a relative of J.C. Adams, the co-discoverer, with Leverrier, of Neptune. This was addressed from Cambridge.

In a booklet on the history of the Royal Observatory, published by H.M. Stationery Office in 1975, the author gives high praise to Sir George Airy, 'greatest Astronomer Royal', whose merits are described as 'flawless'. We are told that a few human failings might have earned him more appreciation although it appears a few pages later that perhaps Airy had one human failing after all. He 'unfortunately' forgot to mention the work of Adams when he wrote a complimentary letter to Leverrier, after reading in June 1846 of Leverrier's calculations concerning the suspected new planet.

Lowell's letter to Turner makes it clear that his own view of Airy's procrastination and short memory was less charitable. The other letter, from the relative of Adams, is couched in more moderate terms, but is equally emphatic. The delay on Airy's part is attributed, rightly or wrongly, to his unwillingness to give precedence to a younger man — Adams at this time was twenty-seven years old and he had begun his researches at the age of twenty-two. The letter accused Airy of vanity and self-interest in contrast to the generous attitude of Adams who 'always sought to give credit to the work of others'. Perhaps, Airy himself, later, may have felt some qualms of conscience and tried to make amends by recommending Adams for the Royal Society's highest honour, the Copley Medal, which was in due course, awarded.

There is always a chance that, even though no new worlds are left for the amateur astronomer to discover, he may in some obscure corner, find useful and interesting scraps of material which can either add to our historical knowledge, or throw a new light on some not insignificant item in the researches of past years.

The following titles are currently in print:

Making and Using a Telescope by P. Moore & H.P. Wilkins (Eyre & Spottiswoode)
Book of the Telescope by A. Frank (Charles Frank)
The Amateur Astronomer and His Telescope by G. Roth (Faber)
Handbook of Telescope Making by N. Howard (Faber)
Constructing an Astronomical Telescope by N. Mathewson (Blackie)
Making Your Own Telescope by A. Thompson (Sky Publishing Corporation)
Amateur Astronomer's Handbook by J.B. Sidgwick (Faber)
Amateur Telescope Making (Vols. I, II, & III) ed. Ingalls (Scientific American)
Telescopes for Sky-gazing by H.E. Paul (Amphoto)
Practical Amateur Astronomy ed. P. Moore (Lutterworth Press)
Astronomical Telescopes & Accessories ed. P. Moore (Lutterworth Press)
Norton's Star Atlas by A.P. Norton (Gall & Inglis)
How To Make a Telescope by Texerau (Interscience Publishing)
Celestial Objects for Common Telescopes (Vols. I & II) (Dover)
Concise Atlas of the Universe by P. Moore (Mitchell Beazley)
Astronomy in Colour by P. Lancaster Brown (Blandford Press)
The Amateur Astonomer by P. Moore (Lutterworth Press)
Beginner's Guide To Astronomical Telescope Making by J. Muirden (Pelham Books)
Astronomy With Binoculars by J. Muirden (Faber)

The following titles are out of print but can be found in some libraries and, occasionally, in secondhand bookshops:

Half-hours With the Telescope by R.A. Proctor
A Field-Book of the Star by W. Olcott
In Star-land With a Three-inch Telescope by W. Olcott
Through My Telescope by W. Hay
Astronomy for Amateurs ed. by J. Oliver
Telescopic Work for Starlight Evenings by W.F. Denning
Astronomy With an Opera-glass by G. Serviss
Pleasures of the Telescope by G. Serviss
Round the Year With the Stars by G. Serviss
The Moon by G. Serviss
Astronomy With the Naked Eye by G. Serviss
The Stars in Our Heaven by P. Lum
Splendour of the Heavens ed. by T.E.R. Phillips & W.H. Steavenson